T0305696

Sustainable Aviation Fuels

Sustainable Aviation Fuels discusses the transfer process of aviation to carbon-neutral flights, including how to manage the transition period. It also demonstrates how to develop and design a new approach for environmentally friendly air transport with reduced emissions.

Covering the full scope of commercialisation, market considerations, advisements on investments and transition challenges of sustainable aviation fuels (SAF), the book tackles questions related to the cost of changing fuel types, competitive market models that can exist parallel to the oil industry and strategies for airlines to implement. It considers reliability requirements for feedstock suppliers and SAF producers, as well as ways to avoid feedstock shortages.

The book will interest aviation industry professionals, fuel producers, airline fuel buyers, airport operators and propulsion engineers working on SAF production. Aviation, aerospace and business students taking courses in propulsion, gas turbine emissions, air transport management, supply chain development and sustainable energy production will find the book useful as well.

Joachim Buse received his PhD from the Institute of Infrastructure and Resources Management at the Faculty of Economics, Leipzig University. He started his professional career in 1986 with Lufthansa German Airlines in various purchasing functions and became Deputy Vice President Fuel Supply in 1992. In 1996, he left Lufthansa and became Managing Director of AFS Aviation Fuel Services GmbH, a joint venture company of Air BP, Shell and Lufthansa. In 2005, Joachim Buse returned to Lufthansa and became CPO of Lufthansa Passenger Airlines. He retired from Lufthansa in 2016 and founded Adeptus Green Management GmbH, a consulting company for Sustainable Aviation Fuels and renewable energy projects. Joachim Buse was a co-founder and board member of aireg – Aviation Initiative for Renewable Energy in Germany in 2011 and moved to aireg's advisory board in 2019. Since 2021, he has supported the HyKero project of EDL Anlagenbau Gesellschaft mbH, Leipzig, dealing with the construction of a 50,000-ton commercial plant for production of PtL-kerosene based on hydrogen and natural carbon sources (e-SAF).

Sustainable Aviation Fuels
Transitioning Towards Green Aviation

Joachim Buse

CRC Press
Taylor & Francis Group
Boca Raton London New York

CRC Press is an imprint of the
Taylor & Francis Group, an **informa** business

Designed cover image: iStock

First edition published 2024
by CRC Press
2385 NW Executive Center Drive, Suite 320, Boca Raton FL 33431

and by CRC Press
4 Park Square, Milton Park, Abingdon, Oxon, OX14 4RN

CRC Press is an imprint of Taylor & Francis Group, LLC

© 2024 Joachim Buse

Library of Congress Cataloging-in-Publication Data
Names: Buse, Joachim, author.
Title: Sustainable aviation fuels : transitioning towards green aviation / Joachim Buse.
Description: First edition. | Boca Raton, FL : CRC Press, 2024. |
Includes bibliographical references and index.
Identifiers: LCCN 2023039286 (print) | LCCN 2023039287 (ebook) |
ISBN 9781032576022 (hardback) | ISBN 9781032576039 (paperback) |
ISBN 9781003440109 (ebook)
Subjects: LCSH: Airplanes—Fuel. | Biomass energy. | Fuel switching. |
Sustainable transportation. | Aeronautics, Commercial—Environmental aspects.
Classification: LCC TL704.7. B87 2024 (print) | LCC TL704.7 (ebook) |
DDC 665.5/3825—dc23/eng/20230927
LC record available at https://lccn.loc.gov/2023039286
LC ebook record available at https://lccn.loc.gov/2023039287

ISBN: 978-1-032-57602-2 (hbk)
ISBN: 978-1-032-57603-9 (pbk)
ISBN: 978-1-003-44010-9 (ebk)

DOI: 10.1201/9781003440109

Typeset in Times
by codeMantra

*This book is dedicated to the pioneers of
sustainable aviation fuels and their contributions
to climate friendly air travel*

*Manfred Aigner
John Benemann
Steve Csonka
John Holladay
Martin Kaltschmitt
Siegfried Knecht
Lourdes Maurice
Klaus Nittinger
Matthew Pearlson
Mark Rumizen
Joachim Szodruch
Nancy Young*

*Once you have experienced flying,
you will walk the earth forever
with your eyes turned heavenwards.*

*For there you have been and there you will
always be drawn back to it.
Leonardo da Vinci*

Contents

Foreword

Commercial aviation's recognition of SAF as a potential means to address environmental constraints on the growth of air travel emerged in the mid-2000s. The initial focus was on the development of the process technologies needed to make drop-in SAFs, along with the approval requirements for use on commercial aircraft. Little thought was given at that time to the challenges of integrating these new types of jet fuels into the existing, highly structured and entrenched jet fuel supply chain. A supply chain was designed around petroleum-derived jet fuel and was managed and controlled to a great extent by the producers of petroleum-derived jet fuel.

While seemingly reasonable accommodations to meet the safety needs of aviation were devised by the aviation fuel community to produce drop-in SAFs, the complexities of integrating the qualified SAFs into the supply chain were not considered. The development of the conversion technologies and the qualification effort were thought to be the major challenges, and the blending requirements and feedstock limitations were thought of as practical and reasonable approaches to ensuring fit-for-purpose jet fuels were produced from the new and novel raw material sources.

As the consequences of unabated carbon emissions become more apparent in today's world, more governmental policy and industry focus is being directed towards the implementation of carbon abatement measures, such as promoting the use of more SAF in aviation. However, with this new focus, stark realities are emerging of what it will take to scale up the use of SAF to a level at which we can reverse or arrest the current carbon emission trends.

In this book, Dr. Buse provides us with an honest portrayal of those realities by systematically breaking down the economic, logistical and policy challenges faced by new SAF producers in comparison to the existing petroleum-derived jet fuel production system and supply chain elements. He also highlights the hesitation by airline management, politicians and regulators to address the challenges posed by these realities and to acknowledge the need to take unprecedented actions that transcend the historical industry and national boundaries that impede global cooperation.

As the more frequent occurrences of wildfires, torrential rainstorms, flash floods and unrelenting heat waves awaken us to the harsh realities of climate change, we now have Dr. Buse's book to guide us in our efforts to address the impact that aviation has on our global environment. The challenges we face to be successful in this task are not insignificant, but we should be inspired to work harder and smarter upon consulting his book.

Mark Rumizen
September 2023

Executive Summary

Civil aviation now faces its biggest challenge in more than 70 years, following the end of the Covid19 restrictions on international aviation: Decarbonising the aviation sector to meet its own self-imposed net-zero emissions target by 2050. Progressive climate change threatens our future and the economic activity of all sectors across borders, countries and continents. Global trade, tourism and even family relationships are inconceivable in the global age without functioning air transport. The damage to the global economy and the hardship suffered by people when aircraft are grounded were brought home to all concerned by the Covid19 pandemic.

As an indispensable means of transportation, aircraft contribute to climate change with their combustion engines and high-altitude exhaust gases, which amplify the greenhouse gas effect that contributes to rising temperatures and the resulting damage to the climate. The aviation sector needs a special jet fuel to power all commercial aircraft: JET A-1 is the fuel optimally adapted to jet engines, used by all airlines worldwide as a unique energy source with high energy density and moderate fuel weight. The rapid development of air transport would not have been possible without this fuel. And here lies the biggest problem facing aviation today: as a fossil fuel derived from crude oil, Jet A-1 contributes negatively to the accumulation of carbon dioxide in the atmosphere, better known as CO_2.

Fortunately, there is a solution to the problem: replace the fossil fuel with an equivalent fuel that, as part of a circular economy, does not cause additional pollution but can eliminate up to 90% of the emissions: SAF – Sustainable Aviation Fuel – is the umbrella term for all new jet fuels that can build on the success story of JET A-1 while minimising aviation's greenhouse gas impact.

This book shows how aviation can make the gradual transition from its fossil fuel past to a climate-friendly future by transforming the sector towards Green Aviation.

The rapid decarbonisation of aviation is an unprecedented undertaking for which there is no master plan. As demand continues to grow, the problem is not cutthroat competition, which incumbents do not fear, but the complexity of the project, with uninterrupted supply to airlines while integrating new products and the indispensable cooperation of existing market players. In this confusing situation, which is akin to flying blind in the fog, this book provides guidance, illustrates the interrelationships and explains the practices and processes in the jet fuel market. New companies are entering the market with innovative technologies and sustainable types of aviation fuel, but without any knowledge of how the market works and the rules accepted by the industry in this niche business.

The future of aviation is at stake. Many people use airplanes but don't know how the aviation industry is organised and operates. That is why the book begins with the reader boarding the plane and getting a glimpse of air transportation – the safety procedure as you taxi to the runway, so to speak. While waiting for take-off clearance, the book gives a brief history of the oil market, whose products are still essential to the transportation sector. The two different industries – aviation and energy – take the reader on a journey through time with the transition from fossil fuels to environmentally friendly aviation fuels. Our journey begins with a comprehensive look at the aviation fuel market. The established market rules for the distribution of aviation fuels have evolved and proven themselves over decades. They are, in a sense, the air routes on which the fuel supply business flies to its destination airports, with the next aircraft refuelling immediately behind. Reliability is a fundamental principle of aviation. The engine only turns if all the wheels mesh. It takes teamwork to make it run like clockwork. The new SAF – Sustainable Aviation Fuels – are the new blade rings in the engine that will ensure a better climate in the combustion chamber.

The transition to renewable fuels requires billions of dollars of investment in new production facilities and the development of new technologies, as well as strategic considerations on how to profitably allocate these investments in the refining sector. In this "race against time", decisions have to be made under uncertainty, especially when there are several alternatives to choose from and the future development of the market cannot be predicted with certainty due to a lack of benchmarks. To this end, the book provides a radar screen showing the areas of stormy weather that an investor should avoid in order to minimise turbulence. Investors in this new market segment lack swarm intelligence because their competitors' moves are made under the same uncertainty as their own decisions. Copying a competitor's strategy is therefore no guarantee of success.

The new renewable fuels also give airlines new options for sourcing the fuels they need. The opportunities, risks and consequences of vertical integration of fuel supply are presented and assessed.

Just as every flight is monitored by air traffic controllers, international bodies and national legislators regulate the sustainability requirements for renewable jet fuels, define the characteristics of the required feedstocks and regulate the market through incentives ("carrots") or sanctions ("sticks"). In this corridor, governments can sustainably promote the positive forces of accelerated market uptake, or they can stifle market forces by imposing restrictions. The complexity of the issue is greatly increased when governments choose different – and sometimes conflicting – policies rather than agreeing on the right trajectory. Current regulations are explained to help decision-makers make their choices.

In the public debate about new aviation fuels and new technologies, one aspect is largely neglected: the raw materials used to produce SAF are an essential, sometimes even decisive factor for or against a technology, a location or an investment decision. Aviation feedstocks are not "just there when you need them". They are subject to strict cultivation rules, must be certified and have their own market laws. Sourcing raw materials for SAF is like flocks of birds all looking for food in the same lake. If the supply of raw materials does not grow in line with SAF's production capacity, a shortage of supply will lead to rising prices in the long term. Unlike a pilot, the book does not shy away from the subject, but illustrates the connections.

It is good practice for the captain to pass on information to the passengers during the cruise flight. In the same way, the book explains the essential aspects of SAF production and certification, the interrelationships to be observed when blending SAF and JET A-1, and the necessary interrelationships for the emission crediting of purchased SAF quantities in the US subsidy regimes and for crediting in the European Emissions Trading Scheme. In the same way that passengers are advised to put their seats in the upright position and to open their window blinds at the beginning of the descent, the book also explains the quality assurance measures that must be observed in the production, transport and storage of SAF, as well as the rather complex liability and insurance requirements in the event of incidents caused by the quality of the fuel or the refuelling of the aircraft during operation.

Finally, eight different strategic approaches are presented that support the market introduction of sustainable aviation fuels in very different ways and enable the goal of rapid decarbonisation of aviation through an instrumented approach to the net-zero runway. That is the goal of this book, and when you feel an inner voice saying – just like arrival of your flight at the airport terminal gate, "All doors in park, please", you will not only have reached the end of your journey, but you will have gained a holistic understanding of the interrelationships in the aviation and oil markets.

List of Figures

List of Tables

List of Acronyms

ABtL	Advanced Biomass to Liquids (FT-SPK and ATJ)
AIAA	American Institute of Aeronautics and Astronautics
Aireg e.V.	Aviation Initiative for renewable Energy in Germany e.V.
AMS	IATA code airport Amsterdam
ASK	Available Seat Kilometres (SKO)
ASTM	American Society for Testing and Materials
ATA	American Transport Association (new designation A4A)
ATAG	Air Transport Action Group
ATJ-SPA	Alcohol to Jet fuel – Synthetic Paraffinic Kerosene with Aromatics
ATJ-SPK	Alcohol to Jet fuel – Synthetic Paraffinic Kerosene
ATL	IATA code airport Atlanta, USA
ATM	Air Traffic Management
AUH	IATA code airport Abu Dhabi, UAE
AVGAS	Aviation Gasoline (aviation fuel for piston engines)
A4A	Airlines for America (Industry Association, formerly ATA)
BAU	Business as usual
BCG	Boston Consulting Group
BER	IATA code airport Berlin
BioSPK	Bio-synthetic paraffinic kerosene
BMDV	Bundesministerium für Verkehr und digitale Infrastruktur (German Federal Ministry of Transport and Digitalization)
BP	British Petroleum plc.
BtL	Biomass to Liquid (process for producing fuel from biomass, FT-SPK)
B2B	Business-to-Business (business relationships with business customers)
B2C	Business-to-consumer (business relations with private customers)
CAAFI	Commercial Aviation Alternative Fuel Initiative (in the USA)
CCS	Carbon Capture and Sequestration
CCUS	Carbon Capture, Utilisation and Sequestration
CDG	IATA code for Charles de Gaulle airport, Paris
CDM	Clean Development Mechanism (certificates in emission trading)
CEO	Chief Executive Officer
CFO	Chief Financial Officer
CGK	IATA airport code Jakarta, Indonesia
CHJ	Catalytic hydrothermolysis jet fuel
CLT	IATA code airport Charlotte, USA
CNG	Compressed Natural Gas
CO	Carbon Monoxide
COO	Chief Operating Officer
CO$_2$	Carbon Dioxide

CPH	IATA code airport Copenhagen, Denmark
CSR	Corporate Social Responsibility
CtL	Coal to Liquid (process for producing fuel from coal)
DBFZ	German Biomass Research Centre (Leipzig, Germany)
DEW	Dry Empty Weight (aircraft empty weight without fuels)
DG MOVE	Directorate-General for Transport and Mobility of the EU Commission
DLR	Deutsches Zentrum für Luft- und Raumfahrt e.V. (German Center for Aeronautics and Aerospace)
DOW	Dry Operating Weight (weight of the aircraft with operating materials and crew, but without aircraft fuel that can be flown out)
dLUC	Direct Land Use Change
DSHC	Direct Sugars to Hydrocarbons (new term "SIP")
EASA	European Aviation and Space Agency (Cologne, Germany)
EEA	European Economic Area (of the EU27 member states)
EEC	European Economic Community (founded 1957 by six countries, later transferred to and integrated in EU27)
EEG	law for the expansion of renewable energies (in Germany)
EDF	Environmental Defense Fund (NGO)
EI	Energy Institute (London, UK)
EIA	US Energy Information Administration
EMCS	Excise Movement Control System (EU customs control system)
eq	equivalent
eSAF	electricity-based Sustainable Aviation Fuel (PtL Kerosene)
ESG	Environmental, Social, Governance (framework of corporate sustainability plans)
ETOPS	Extended Twin Operation
ETS	European Trading Scheme
EU	European Union
EU-OPS	EU Directive (EC) No 859/2008 on minimum safety requirements for aircraft for Aircraft Supplementary to Directive (EEC) No 3922/91
EU-RED	European Union – Renewable Energy Directive
EUR	Euro
FAA	Federal Aviation Administration
FAME	Fatty Acid Methyl Ester
FAR	Federal Aviation Regulation
FAO	Food and Agriculture Organization (UN agency, Rome, Italy)
FDF	IATA code airport Fort-de-France, Martinique
FGP	First Gathering Point (transport/processing collection point)
FORDEC	Facts/Options/Risks/Decision/Execution/Control (acronym in aviation as part of Crew Resource Management)
FRL	Fuel Readiness Level (CAAFI)
FT	Fischer–Tropsch (process for synthesising gas streams)
FT-SPK	Fischer–Tropsch Synthetic Paraffinic Kerosene
FT-SKA	Fischer–Tropsch Synthesised Kerosene with Aromatics

FTC	Fuel to Carry
GHG	Greenhouse gas emissions
GMBM	Global Market Based Measures (CO_2 compensation system)
GREET	Greenhouse gases, Regulated Emissions and Energy use in Transportation (developed by Argonne National Lab, USA)
GtL	Gas to Liquid (process for producing fuel from gas)
HAM	IATA airport code Hamburg, Germany
HC-HEFA	Synthesised Paraffinic Kerosene from Hydrocarbon – Hydroprocessed Esters and Fatty Acids
HEFA	Hydroprocessed Esters and Fatty Acids
HRJ	Hydrotreated Renewable Jet (-Fuel)
HVO	Hydrogenated Vegetable Oil
IATA	International Air Traffic Association
ICCT	International Council for Clean Transportation (Washington D.C., USA)
ICAO	International Civil Aviation Organisation
ICSA	International Coalition for Sustainable Action
IEA	International Energy Agency
IFQP	International Fuel Quality Pool (of IATA)
ILUC	Indirect Land Use Change
ISCC	International Sustainability and Carbon Certification (Cologne)
IT	Information Technology
IZF	Internal rate of return methodology
JET A	Type designation for specified jet fuel
JET A-1	Type designation for specified jet fuel
JIG	Joint Inspection Group (petroleum industry inspection group)
JG	Joint Guidelines
kg	Kilogram
kt/a	Kilo tons per annum (unit of measurement in 1,000 tons per year)
KUL	IATA airport code Kuala Lumpur, Malaysia
kWh	Kilowatt hours
L	Litre
LCA	Life Cycle Analysis
LCFS	Low Carbon Fuel Standard (Cal., USA)
LTAG	Long-Term Aspirational Goal (ICAO)
LUC	Land Use Change (direct land use change)
MAD	IATA airport code Madrid, Spain
MBM	Market-Based Measures (CO_2 compensation system)
MCO	IATA code airport Orlando, Florida, USA
MJ	Mega joule
MMBtu	Million British Thermal Unit (equals to 1,055 MJ or 293 kWh)
MJ	Mega Joule
MOB	IATA code airport Mobile, Alabama, USA
mt	metric ton (1,000 kg)
MSW	Municipal Solid Waste
MTOW	Maximum Take-off Weight

MZFW	Maximum Zero Fuel Weight (max. structural weight with passengers and cargo
NexBtL	Next (Generation) Biomass to Liquid (Fuel)
NGO	Non-Governmental Organisation
NM	Nautical Miles
n.a.	Data not available
OECD	Organisation for Economic Co-operation and Development (Paris)
O&D	Origin & Destination (Traffic)
PCB	Polychloride Biphenyl
PCPP	Polychloride Phenoxyphenol
PRG	IATA code airport Prague, Czech Republic
PtL	Power to Liquid (jet fuel from hydrogen via electrolysis)
RFNBO	Renewable Fuel of Non-Biological Origin
RPK	Revenue Passenger Kilometres (SKT)
ROI	Return on Investment
ROIC	Return on Invested Capital
RSB	Roundtable for Sustainable Biomaterials (Lausanne, Switzerland)
SAF	Sustainable Aviation Fuels
SBC	Synthetic Blending Component (definition by EI)
SBTi	Science-Based Targets (initiative)
SESAR	Single European Sky ATM Research project
SGF	Sectoral Growth Factor (ICAO CORSIA)
SIP	Synthesised Iso-paraffin (product name "Farnesane")
SJU	IATA code airport San Juan, Puerto Rico
SLF	Seat Load Factor
SKO	Seat Kilometres Offered
SKT	Seat Kilometres Taken
STR	IATA code airport Stuttgart, Germany
SXM	IATA code airport St. Maarten, Leeward Islands
t	Time; ton, ton
T	Ton
TBO	Trajectory-Based Operations
To	Ton
TKO	Ton Kilometre Offered
TOW	Take-off Weight (take-off weight of an aircraft)
TUN	IATA code airport Tunis, Tunisia
UAE	United Arab Emirates
UKP	United Kingdom Pounds (Currency)
US	United States
USA	United States of America
USC	United States Cent
USD	United States Dollar
US$	United States Dollar
USG	United States Gallon
USP	Unique Selling Proposition
UT	Upper Troposphere

VUCA	Volatility, Uncertainty, Complexity, Ambiguity (acronym for complex framework conditions in corporate management)
WAW	IATA code airport Warsaw, Poland
WEF	World Economic Forum
WtL	Waste to Liquid (FT-SPK)
WWF	World Wide Fund for Nature (NGO, based in Gland, Switzerland)
ZFW	Zero Fuel Weight (aircraft and payload weight without fuel)

1 Global Aviation – A Forward Looking Overview

Global mobility and the movement of goods require aviation as the fastest means of transport over long distances. In 2019 – the last financial year not affected by the Covid19 pandemic –4.56 billion passengers were transported in passenger aircraft, corresponding to 8,686 billion revenue passenger kilometres (RPKs), resulting in an average load factor of 82.4% of the sum of available seat kilometres. A total of 57.6 million tons of freight were transported in the belly of passenger aircraft or on dedicated freighters in 2019, totalling 391 billion freight ton-kilometres. Compared to 2018, global passenger traffic increased by 4.9% and global air freight by 6.8% (ICAO, 2020).

Domestic and international air transport suffered from the decline in passenger and freight demand caused by the Covid19 pandemic and the associated flight restrictions and reductions from 2020 to mid-2022. However, air transport is in a promising recovery phase and will continue to grow beyond the 2019 figures in 2024 (IATA, 2022).

For the prosperity of international trade and global passenger travel, aviation is the cornerstone of efficient transport and represents the modern way of life. For perishable goods from abroad, there is no alternative to air transport. Industrial production depends on just-in-time deliveries of equipment, consumer goods and spare parts. Disaster relief, such as earthquakes, floods or storms, requires immediate action to rescue people and provide medical care, food and shelter, which air transport can provide. Business travel enables international sales and distribution of products, ethnic travel is a consequence of global education and international job opportunities for skilled people who wish to keep in touch with their families and relatives. Leisure travel is a lifestyle choice for individuals, couples and families with good incomes and work/life balance opportunities in modern societies.

Air travel is the most time-efficient mode of transport over long distances, and travellers to distant islands or other continents usually have no choice but to fly, unless they wish to take a ferry or cruise, where travel time is not an issue.

Even if individuals do not actively choose to travel by air, life today is supported by air transport in many ways that go beyond any subjective perception.

From an environmental point of view, air transport does not require the large infrastructure that roads and railways do. Flying from point to point requires only airport terminals, aprons and runways at both ends, and some navigational radar and some directional and non-directional beacons along the airways. Land use change is limited to airports, terminals and runways. There are almost no emissions associated

DOI: 10.1201/9781003440109-1

with the maintenance and replacement of aviation-related infrastructure compared to roads and railways. However, aircraft burn jet fuel in their engines while flying. The environmental impact of aviation is concentrated in engine exhaust and, due to physical constraints and the outstanding research and development of alternative energy sources, the burning of jet fuel in jet engines will remain the predominant source of energy for aircraft until 2050 and beyond, and due to the growth in domestic and international air travel, the increasing consumption of fossil jet fuel – despite efforts to reduce specific engine fuel consumption – contributes to an increase in the so-called greenhouse gases, which are the main cause of global warming.

Replacing fossil jet fuel with new environmentally friendly types of jet fuel is not a matter of choice for aviation, it is a necessity. The new generation of environmentally friendly aviation fuels began in 2009 with the approval of the first synthetic paraffinic kerosene by the American Society for Testing and Materials (ASTM), now ASTM International. The name "SAF" as a generic term for all sustainable aviation fuels, regardless of the nature of the raw materials and production processes, only became popular since 2017 and is now a globally accepted term.

1.1 GLOBAL AVIATION – AN INTEGRAL VIEW (GROWTH PATTERN, FLEET SIZE AND AIRCRAFT REPLACEMENTS)

In the aftermath of the Second World War, the nations affected by military action agreed to create a supranational institution to deal with the existing and future challenges of global conflict by founding the United Nations (UN) on 24 October 1945 in San Francisco, with its headquarters in New York, USA. A year earlier, on 7 December 1944, the International Civil Aviation Organisation (ICAO) had been established in Chicago as an institution to harmonise global aviation standards. On 13 May 1947, ICAO became an official sub-organisation of the UN, with headquarters in Montreal, Canada. In addition to standardising air traffic regulations, one of ICAO's main tasks was and still is to promote international aviation as an instrument of peace. As long as people from different cultures and regions can come together and understand local and national customs, aviation can contribute to the peaceful and friendly coexistence of nations and help to mitigate conflicts. As a result, ICAO decided that Member States must continue to exempt jet fuel for international flights from all fuel-related taxes. This rule is still in force and is binding. However, due to difficulties in adequately distinguishing between jet fuel consumed on domestic and international flights, many countries have also exempted domestic jet fuel from taxation. To date, ICAO has made an unparalleled contribution to the rapid development of aviation: In 2019, a total of 1,478 airlines operated 33,299 commercial aircraft and connected 3,780 airports with scheduled commercial flights. A total of 48,044 routes were served. Around 50% of these routes are point-to-point or city pairs. 46.8 million scheduled commercial flights were counted in 2019. In terms of passengers carried, 58.4% (or about 2.6 billion) were on domestic routes and 41.6% (or about 1.9 billion) were on international routes (ATAG, 2020) (Figure 1.1).

Global aviation contributed USD 3.5 trillion to GDP in 2019, or about 4.1% of global economic activity. If global aviation were a country, its GDP would rank 17th, similar to Indonesia or the Netherlands. Commercial airlines employ 3.6 million people.

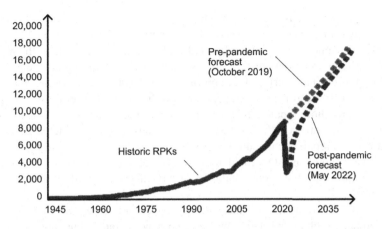

FIGURE 1.1 Global air traffic historic revenue passenger kilometres (RPK) and forecast in billion RPK per year.

Source: IATA Economics/Tourism Economics (IATA, 2022) (prepared by author).

TABLE 1.1
Available Seat Kilometres (in millions) and Increase in Passenger Load Factor (in %)

Year	Passenger-km [millions]	Available Seat-km [millions]	Load Factor [%]
2010	4,930,250	6,307,072	78
2011	5,254,557	6,736,041	78
2012	5,535,641	7,019,380	79
2013	5,839,696	7,347,189	79
2014	6,188,735	7,763,235	80
2015	6,652,791	8,291,255	80
2016	7,144,498	8,898,862	80
2017	7,716,542	9,470,759	81
2018	8,278,782	10,124,456	82
2019	8,685,667	10,540,946	82

Source: ICAO (2023) (prepared by author).

Airports and airport service providers employ a further 6.1 million people to handle passengers and air cargo (ATAG, 2020) (Table 1.1).

Aircraft are an airline's most important asset. The assembly of an aircraft, including third-party equipment such as engines, cockpit instruments, landing gear, cabin seats and galleys, takes approximately 12–18 months from the start of assembly to the final assembly and roll-out of the new aircraft. The normal life of an aircraft is estimated to be around 30 years. Actually, the aircraft life is determined by the cumulative number of flight hours flown and the cumulative number of flight cycles performed. The latter reflects the fatigue impact of each flight, as the airframe is pressurised when flying above 6,000 feet or 2,000 m, and such pressurisation causes cracking and deformation of the fuselage if the aircraft is operated beyond its guaranteed life and performance limits. Due to the average route length, narrow-body aircraft are certified for over 45,000 total cycles, while wide-body aircraft are certified for approximately 30,000+ cycles. Due to the high investment, an aircraft is usually operated until the end of its useful life. As ageing aircraft require more repairs and associated downtime, legacy carriers are looking to sell older aircraft to other operators well in advance of an increase in unexpected critical repairs that require the aircraft to be grounded while repairs are completed – known as AOG incidents – resulting in the rebooking of passengers and, in some cases, the accommodation of passengers until alternative transport can be provided.

The useful life of an aircraft has a direct impact on fleet renewal. Technological improvements, such as lower fuel burn due to new engine technology, take a long time to replace a significant proportion of aircraft with more efficient replacements. If the size of the market remained constant and aircraft were only replaced as they aged, it would take 35–45 years for an aircraft model to be completely replaced by its successor across all operators of that aircraft type. As demand for air travel continues to grow, fleet expansion is the driver for additional aircraft orders, as well as the replacement of existing aircraft with newly developed models that offer improvements in operating and maintenance costs, transport capacity, seating capacity and range. Based on an average market growth of 4%–6% per year, civil aviation doubles its transport capacity every 17 years on average. This growth is reflected in the increasing size of airframes with higher seating capacity and the number of aircraft in service. In general, the replacement of ageing aircraft and the expansion of the fleet due to growth will result in new aircraft versions of existing models with lower fuel consumption and thus extended range or increased payload, or in a completely new aircraft design. It typically takes 5–7 years to design, engineer, build, test and certify a new aircraft type (Figure 1.2).

While aircraft deliveries moved up to more than 1,800 new commercial aircraft in 2018, the aviation industry's profitability already deteriorated from its best rate of 8.5% ROIC in 2014 to 7.5% ROIC in 2018, reflecting a tense pricing situation in the passenger and cargo markets.

1.2 JET FUEL AS THE SINGLE SOURCE OF ENERGY FOR AIRCRAFT

Today's jet fuel is the result of military developments during the Second World War with the invention of the first generation of jet engines for fighter aircraft. The first military jet fuel was certified as JP-1 (Jet Propellant 1), which contained gasoline, kerosene and other chemicals.

Its thermal and storage stability made aviation turbine fuel (the legal term for jet fuel) the perfect product for civil aviation, limiting the risk of quality degradation during transport or storage.

The introduction of jet engines in passenger aircraft in the early 1960s enabled airlines to fly at higher speeds and cruise at higher altitudes than with the piston engines that had dominated aircraft design since the Second World War. The

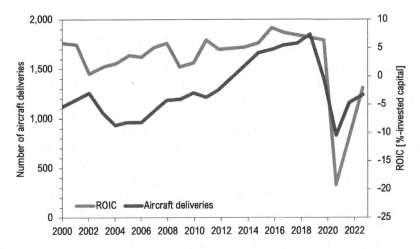

FIGURE 1.2   Aircraft deliveries per year and airline industry ROIC.

Source: IATA (2022) (prepared by author).

traffic growth of the 1960s and the new engine types required a new type of fuel that combined several requirements in one specification: (1) high energy density at a reasonable fuel weight, (2) a freezing point of −47° on the centigrade scale to allow flight levels up to 40,000 feet above sea level, (3) a high flash point and good vaporisation in the engine combustion chamber, and (4) availability at all airports around the world.

Finally, aviation jet fuel was consolidated into a harmonised specification for a light petroleum product under the grade name "JET A-1". JET A is the primary fuel used in the US, while JET A-1 is the primary fuel used in other regions of the world. These are basically the same fuel, differing only in the maximum allowable freezing point. Jet A-1 fuel has been approved by US ASTM D1655 and British Def. Stan. 91-091 and Canadian CGSB 3.23. The specifications describe fuel properties with minimum or maximum values or ranges. Consequently, there is no single production method for refining kerosene from crude oil. Based on different refining technologies and different crude oil sources, the chemical composition of each batch of JET A-1 will vary due to differences in crude oil sources and refining techniques. JET A-1 also typically requires the use of approved additives necessary to meet specification requirements, such as antioxidants, metal deactivators, corrosion inhibitors and static dissipator additives.

The density of JET A-1 is standardised between 0.775 kg/L (minimum density) and 0.84 kg/L (maximum density). The density ratio reflects the energy content of the kerosene and determines the fuel quantity calculation for a flight, as a low-density jet fuel results in higher fuel consumption during the flight and this must be taken into account in the fuel quantity calculation when an aircraft has to be refuelled for the next flight.

The new generation of aircraft powered by jet engines revolutionised the airline business: The improved cruise speed of jet engines allowed pilots to fly at higher altitudes with less drag and less turbulence requiring speed reductions, evasive manoeuvres or even contingency flight plans. The lower air density had a positive effect on fuel consumption and, together with improved engine performance, long-haul aircraft were able to fly long distances without technical landings for refuelling. The huge increase in international routes required the availability of JET A-1 and identical fuel types around the globe, at all times and with the same production quality.

Aircraft manufactures specify what type of aviation turbine fuel must be used on their aircraft (universally this is Jet A and Jet A-1) and the regulatory authorities require airlines to only use the specified fuel. However, because the fuel is not a permanent part of the aircraft, the production, distribution and storage of JET A-1 in the supply chain upstream of the aircraft is not regulated by governments or multinational bodies. As a result, there are no government regulations governing fuel quality, quality inspections and routine administrative quality monitoring. To address this situation, major oil companies have voluntarily formed a Joint Inspection Group (JIG). Under JIG, engineers from different oil companies have drawn up recommended standards for the production and handling of jet fuel. To date, JIG members from various oil companies organise a schedule of inspections of refineries, airport fuel depots and airport fuelling services to be carried out every 6 months at all jet fuel related facilities to monitor compliance with these standards. Such inspections are non-competitive as the common task of all inspectors is to ensure the highest possible level of product quality, operational safety and avoidance of environmental incidents.

The standards issued by JIG are frequently updated and contain all the rules and regulations to be observed by all companies along the jet fuel supply chain.

Following EU Directive 965/2012, IATA member airlines also started a voluntary technical inspection group called IFQP (IATA Fuel Quality Pool). Group members share the results of technical inspections of fuelling equipment and storage facilities at airports. Both inspection groups work closely together to promote fuel quality and safety.

High aircraft utilisation requires the maximum number of flights per day within given constraints, such as airport operating hours. Aircraft refuelling between flights must take place within a specified time frame after passengers have left the aircraft cabin and before new passengers board. At large airports with busy gate positions and long-haul flights requiring large quantities of jet fuel, aircraft refuelling is carried out using an underground hydrant system connected by pipelines to the airport jet fuel depot, which operates under commingled storage conditions. All deliveries of jet fuel from different sources meeting the JET A-1 specification can be stored together, as the same blending applies to aircraft fuel tanks containing residual fuel from previous flights. A hydrant system is expensive to install and maintain, and it only pays for itself if large volumes are dispensed through the airport's pipeline network. Aircraft refuelling at small- and medium-sized airports is carried out by bowser trucks, which are loaded at the airport jet fuel depot and pump jet fuel from their vehicle-mounted tank to the aircraft. The efficiency of today's aircraft utilisation is based on the continuous development of aircraft and equipment, operational procedures and a skilled workforce that performs all tasks to a professional standard.

As long as aircraft require jet fuel as their sole source of energy, the jet fuel supply chain will undoubtedly be the backbone of modern aviation.

Airlines report various statistics on their transport capacity. However, data on fuel consumption is rarely available. Due to the existing tax exemption on kerosene used for commercial flights, national governments also need to have an overview of annual kerosene sales at all airports in their countries, and such data could easily be aggregated into a global database. In fact, there is no clear measure of global jet fuel consumption. One ton of jet fuel emits 3.15 tons of CO_2 when burned in an aircraft engine, reflecting a fixed relationship between fuel and emissions. Fuel and emissions data vary when looking for exact jet fuel consumption for 2019. Most publications refer to the CO_2 emissions produced without specifying the fuel consumption base. In all likelihood, the total jet fuel consumption of civil aviation in 2019 was 290 million tons of jet fuel, resulting in 913.5 million tons of CO_2. The share of transport emissions has been published as 24.45% of total emissions from all sectors (Statista, 2022). The share of aviation in total emissions varies between 1.9% and 3.06%, depending on the data source (Statista, 2022). Neither ICAO nor IATA publish actual jet fuel uplift data, which would be extremely helpful in determining future demand up to 2050, which can be derived from an estimated dynamic growth pattern. Assuming that the base volume of 290 million tons of jet fuel (or 362.5 million cubic metres or 95.762 billion US gallons at an average density of 0.8 kg/L) is a good match for actual consumption in 2019, a net dynamic growth in jet fuel demand of 2% per year would result in a demand size of 535.8 million tons in 2050 (+84%). Cumulative jet fuel consumption between 2019 and 2050 would total 12.859 billion tons of jet fuel, resulting in cumulative emissions of 40.401 billion tons of CO_2 (Figure 1.3).

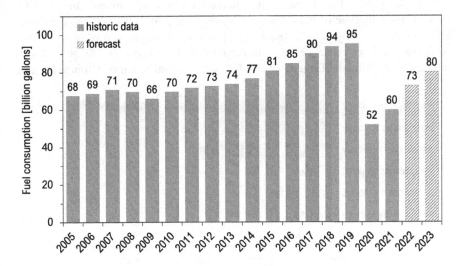

FIGURE 1.3 Total fuel consumption of commercial airlines between 2005 and 2021 with forecast for 2022 and 2023.

Data source: IATA, S&P Global Platts (prepared by author).

1.3 AVIATION AND THE IMPACT ON CLIMATE (EXHAUST EMISSIONS, FLIGHT ROUTINGS, FLIGHT LEVELS, ETC.)

The combustion of fossil jet fuel in aircraft engines is the main source of environmental damage caused by aviation. Since the beginning of the so-called "industrial age", energy has been produced from fossil sources such as coal, oil and gas. Energy conversion in the transport sector began with the combustion of coal and oil in steam engines to convert energy into motion in ships and locomotives, followed by the direct combustion of oil in car and truck engines. The first powered flight by the Wright brothers in 1903 marked the beginning of commercial aviation. However, since this historic milestone, aviation has seen a steady increase in air traffic and, with it, the emissions associated with the combustion of fuel in aircraft engines (Figure 1.4).

The pollution of the atmosphere by exhaust gases has doubled since the beginning of the "industrial age" to 421 ppm CO_2 today and the problems resulting from the increasing emissions of greenhouse gases have led to global warming with adverse effects on our planet. Carbon dioxide (CO_2) has been identified as the main cause of global warming. However, nitrogen oxides (NO_x) and sulphur compounds such as sulphur oxides (SO_x) and dihydrogen sulphate (H_2SO_4), collectively known as "non-CO_2 emissions", have also been identified as greenhouse gases. Soot is a combustion residue that also appears as carbon in jet fuel combustion. Soot particles combine with water vapour to form an exhaust gas that is the basis for the formation of ice crystals surrounding a soot particle in dry air at ambient temperatures prevailing at high altitudes, causing contrail formation. A third component of aviation's impact on the climate is related to high-altitude flights. While flights at flight level 300 or below (30,000 feet, troposphere) still allow some air circulation, flights in the so-called tropopause ($\leq$40,000 feet) operate in a non-convective environment, and the CO_2 emitted has a lifetime of more than 200 years (DLR, 2021). Damage to the ozone layer, which normally protects our atmosphere from being heated by sunlight, has been identified as a consequence of air pollution (IPCC, 1999). The Kyoto Protocol, initiated in 1997 by the United Nations Framework Convention on Climate Change

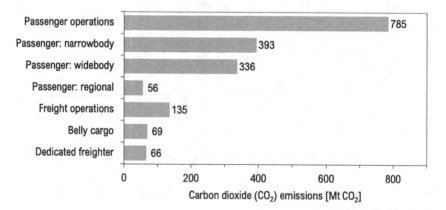

FIGURE 1.4 Carbon dioxide emissions from commercial aviation in 2019 by operation in million metric tons.

Data source: ICCT; Statista 2023 (prepared by author).

(UNFCCC), was concluded as an international agreement to limit greenhouse gas concentrations to a level that would avoid anthropogenic interference with the global climate system. Although not all UN member states signed the Kyoto Protocol, the agreement marks the beginning of climate change mitigation. At the time, the expected goal was to limit global warming to a maximum temperature increase of +2.0°C. Now, faced with the effects of global warming on the world's weather through storms, floods and damage to forests and agriculture, the global ambition to slow further temperature increases has led to the so-called "Paris Agreement", a legally binding international treaty on climate change, which was adopted by 196 nations in Paris, France, on 12 December 2015. This agreement requires all signatories to make all reasonable efforts to limit the global temperature increase to +1.5° Celsius (UN, 2016) (Figure 1.5).

While policymakers have begun to actively seek solutions to mitigate the climate impact of aviation through regulation, incentives and research programmes, the response from airlines to actively reduce their emissions has been relatively negligible. In the wake of high crude oil prices, airframe and engine manufacturers have begun to develop new technologies that reduce fuel burn, which in turn reduces aircraft operating costs. Lower fuel burn automatically results in lower emissions. Therefore, many airlines still argue that their contribution to climate change mitigation is to purchase new aircraft that emit less greenhouse gases and noise. In effect, airlines are doing what they have been doing for the past 50 years: Replacing end-of-life aircraft with new models and adding aircraft to their fleets to keep up with market growth. In addition, the latest development in geared turbofan engines generates more thrust through a gearbox that allows the fan to rotate at a higher speed, resulting in more bypass air and fuel burn improvements of up to 10%. As long as such aircraft operate on existing routes, the geared turbofan – currently only available for narrow-body aircraft – will reduce fuel burn. However, with the addition of a rear centre tank, aircraft such as the A321XLR can fly long-haul routes of up to

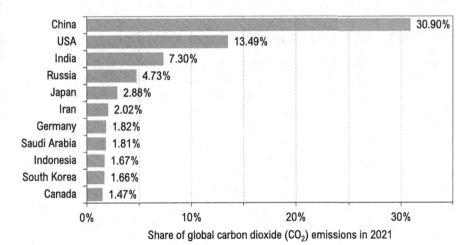

FIGURE 1.5   Share of worldwide CO_2 emissions Top 11 countries by emissions percentage.

Data source: Global Carbon Project; Statista 2023 (prepared by author).

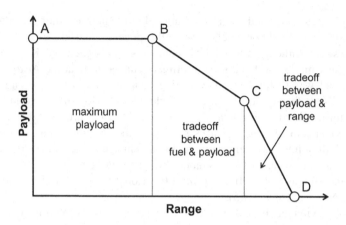

FIGURE 1.6 Aircraft Payload range diagram (illustrative) (prepared by author). Explanation: AB, range distance not affecting payload and MTOW; BC, total flight distance requires payload reduction to MTOW; CD, flight distance requires further reduction of payload at given fuel quantity to gain more range (reduced MTOW).

8,700 km or 4,700 miles in approximately 11 hours. The efficiency gain in fuel burn – theoretically up to 30% per seat – is reversed when such aircraft fly longer routes with a total fuel burn higher than the fuel burn on the conventional routes they previously flew. The development of the geared turbofan aircraft is preparing the market for new routes that are not currently served due to insufficient load factors. These long-haul narrow-bodies will also find their niche in connecting secondary markets with non-stop flights rather than hub-based multi-leg connections. This new capacity will drive significant growth, potentially leading to increased demand for jet fuel and higher emissions before effective emissions reductions are achieved through SAF mandates (Figure 1.6).

1.4   ICAO – INTERNATIONAL RULES FOR AVIATION

Founded in 1944, ICAO is the international body for national departments of transport. ICAO's General Assembly meets every 3 years. In the intervening period, a Council of representatives from 36 Member States, elected by the General Assembly for 3-year terms, is the decision-making body. ICAO has set global standards for airports, air routes, aircraft operations and operational safety. Air traffic management follows ICAO rules and the ICAO-developed concept of "Freedom of the Air" is a globally accepted legal standard for bilateral air traffic agreements between governments that govern the reciprocal rights of airlines operating between two states. In 1983, as environmental concerns in aviation began to grow, ICAO established CAEP, the Committee on Aviation Environmental Protection, to propose policies and regulations to the General Assembly. CAEP consists of 31 representatives from Member States and 21 observers from various associations. CAEP decisions are forwarded to the ICAO Council for final adoption by the General Assembly. These decisions are then transposed by ICAO into the national law of the Member States. The national Department of Transport or Ministry of Transport is the executive body for the transposition of ICAO resolutions into national law or national directives.

1.5 ENVIRONMENTAL OBJECTIVES OF AVIATION

Most environmental improvements in aviation have their roots in protests by dissatisfied citizens outside the aviation industry but directly affected by noise and air pollution around airports. Going back to the beginning of the jet age in the 1960s, the first generation of jet engines were noisier than their piston predecessors. People living near airports protested against aircraft noise, and local authorities were forced to play a mediating role between airports and people living along runway approaches. In addition, first-generation jet engines produced visible exhaust smoke that worsened air quality around airports. Governments responded to protesters' demands for less noise and pollution by requiring detailed measurements of noise and fume distributions and noise mapping. Airports had to compensate citizens living in affected areas for changing window panes to reduce noise, while airports had to restrict the operation of noisy aircraft during the early morning and late evening hours. At the same time, airlines operating noisy aircraft had to pay higher landing fees. As a result, engine and airframe manufacturers have gradually reduced noise and air pollution. New aircraft models produce less noise than their predecessors, and the former "noise carpet" of 4–7 km beyond the runway threshold has been successfully shortened to the airport fence just behind the approach lights.

When early protesters complained about air pollution around airports, local and regional authorities called for regulation of pollution limits. ICAO set up a working group within the CAEP committee to measure and reduce air pollution. As part of the environmental movement in the late 1970s, aviation emissions became an issue for green activists and found its way into the manifests of newly formed green parties in Europe. In 2008, as a first step towards regulating air pollution in Europe, the EU Commission decided that intra-European flights would have to be included in the EU Emissions Trading Scheme ("EU-ETS") from 2013. Since then, airlines have had to offset their CO_2 emissions from intra-European flights (with 85% free allowances in the start-up phase). Again, the initiatives to reduce aviation emissions were taken neither by the airlines nor by the passengers. Several environmental NGOs and green parties in several EU member states called for action, and their insistence on greening society eventually led to new rules for aviation. As the green movement gained momentum, a number of entrepreneurs who had successfully set up new businesses in California, ranging from "garage-size" start-ups to multinational corporations, also embraced the concept of climate-friendly production and distribution of their products, demanding substantial reductions in emissions from their production facilities as well as from all companies involved in their supply chain.

At the same time, any unexpected movement in crude oil prices to unprecedented levels triggered a public debate, driven by the media, about the visible end of the crude oil age. Journalists and oil market pundits predicted that reduced exploration for new oil fields would lead to a reduction in refining capacity and less available volumes of refined oil products, which together would lead to a coming century of high oil prices as a result of an imbalance between supply and demand.

The simultaneous occurrence of peak oil in 2008, the expected reduction in available oil products and the search for climate-friendly improvements in production and transport have triggered the development of the so-called "alternative aviation fuels". Meanwhile, peak oil – the maximum achievable production of crude oil –has lost its threatening potential, and the evolution of oil prices has shown that, in the coming decades, a decline in oil production may be caused by reduced demand in Western societies, but not by a shortage due to limited availability of crude oil. Nevertheless, the threat of oil scarcity and the need for climate-friendly fuels for road transport and aviation have stimulated research and development of non-fossil fuel alternatives by universities, research organisations and companies since the oil price peaked at USD 145/bbl in June 2008.

Despite the steady increase in jet fuel demand associated with growing fleet sizes and increased aircraft utilisation, airlines are not contractually securing jet fuel supplies in line with their medium- to long-term jet fuel requirements. Instead, they are behaving like ordinary consumers who expect to find a petrol station in every town with unlimited supply and the ability to switch suppliers if a competitor offers better prices. Commitments by airlines to begin vertically integrating the supply of fossil jet fuel as part of an integrated supply chain that would reduce jet fuel-related operating costs have been limited to one case in the US. In 2012, Delta Airlines acquired a refinery in Monroe, Pennsylvania that was being prepared for closure. The airline invested $150 million in the purchase and a further $100 million in upgrading the facility. Operated by BP on behalf of the airline, the oil company agreed to exchange all by-products of kerosene production for additional supplies of jet fuel. Like other refineries, Delta's refinery was affected by the Covid19 pandemic, but is expected to return to profitability in 2023 (New York Times, 2022). While other airlines have closely watched Delta's move into refining as an option to reduce their jet fuel costs, no other airline has yet decided to copy Delta's refining approach.

While airlines have come to expect an unlimited supply of jet fuel from their chemical industry partners, they have paid little attention to their environmental footprint. To satisfy their shareholders and passengers, airline annual reports have been expanded to include "CSR reporting" on corporate social responsibility, including all "ESG activities" related to environmental, social and governance issues. EU-based companies with more than 500 employees must comply with the EU regulation on ESG reporting. In the US, the Securities and Exchange Commission is considering an ESG disclosure rule, preferably as part of the IFRS rules. The sustainability section of airline annual reports tends to describe environmental efforts to renew fleets to improve overall fuel consumption and to call for improvements in government-controlled air traffic control that could save flight time and fuel and lead to better utilisation of aircraft. Based on government incentives in the US, US airlines have intensified their efforts to replace fossil A-1 jet fuel by blending Sustainable Aviation Fuels (SAF) with conventional jet fuel. United Airlines and Cathay Pacific have co-invested in Fulcrum's SAF refinery as part of their SAF supply agreements, and United Airlines has established a subsidiary, United Ventures, to seek investment in additional SAF refinery projects to supply climate-friendly jet fuel to United's US network.

In response to the upcoming SAF blending mandates, some airlines have decided to participate in the build-up phase of new SAF production facilities by entering into longer-term off-take agreements to secure SAF volumes that will sooner or later meet their SAF blending mandates.

In fact, alternative propulsion systems, such as battery-powered electric engines or hydrogen-based engines using liquid hydrogen, are still in the early stages of development and will not be available to replace conventional jet aircraft until after 2050. Until the current limitations of battery energy storage and on-board compressed hydrogen storage are overcome, the general conclusion that new technologies will show substantial improvements within a short period of time may be misleading, as technology development may stall when it reaches physical limits that cannot be overcome.

1.6 THE ENERGY TRANSITION CHALLENGE FOR AVIATION

JET A-1 is the ultimate aviation fuel. Not only in terms of its fuel properties, which are perfectly suited to modern jet engine design, but also in terms of a global supply chain that has been optimised over the past 60 years, with numerous refineries, fuel storage depots, hydrant systems, pipelines, rail cars, barges and trucks dedicated to storing or transporting a single product that has proven its reliability and performance characteristics. The design and layout of the latest generation of large turbofan engines is still based on the energy content provided by the JET A-1. Aircraft performance data, maximum take-off weight (MTOW) and therefore the runway length required for safe take-off are based on the energy content of JET A-1. The product itself has not been changed for decades. Some additive suppliers may have developed new additives with improved performance, but the core behaviour of jet fuel has not been affected by such improvements.

Over time, a number of market practices have also shaped industry standards. Even the wording of supply contracts has been harmonised over time, and airline tendering procedures follow voluntary recommendations that are accepted by both buyers and sellers. In the US, airline fuel purchasing departments not only play an active role in concluding fuel purchase agreements but also organise the transport of jet fuel and take ownership when agreeing "ex-refinery" supply contracts. Airline consortia jointly own airport fuel depots in the US and contract third-party service companies to operate these depots and provide aircraft refuelling services.

In Europe, Africa and Asia, airline fuel purchasing departments enter into "free into-plane" contracts, meaning that they take ownership of the fuel when it passes the inlet valve of the fuelling bay on the aircraft wing. The entire supply chain is managed by the selling oil company, including distribution to and storage at airports, as well as third-party aircraft refuelling services on its behalf.

Overall, the JET A-1 supply chain has proven its reliability, product quality assurance and product availability. Supply disruptions are infrequent and always professionally managed, limiting the number of fuel-related flight irregularities to a negligible number. The prevailing jet fuel pricing model is linked to spot market jet fuel prices, resulting in monthly price adjustments and, in the case of free into-plane contracts (which include storage and the required aircraft fuelling service), limited to an airline's actual uplift volume at an airport covered by the airline's fuel supply agreement with a jet fuel supplier. Airlines are affected by upward movements in the price of jet fuel, but with appropriate fuel price hedging contracts, fuel budget stability can be achieved, at least in the medium term, by hedging the fuel price within agreed limits. To summarise the benefits of JET A-1, both delineated fuel supply chain models work like clockwork and have proven their maturity to airlines around the world.

Apart from the environmental impact on the atmosphere caused by burning fossil jet fuel, there is no other reason to replace JET A-1.

Any non-fossil replacement product for JET A-1 must have identical or at least similar fuel characteristics that can be integrated into the existing jet fuel supply chain while maintaining safe aircraft operations at current levels of safety and quality. It should be noted that at many airports jet fuel depots are part of the centralised infrastructure of an airport, with blending of fuel batches delivered to the depot. In addition to the investment required to build an alternative storage facility, most airports are not in a position to provide additional space for a second airport fuel depot and the necessary aircraft refuelling equipment at each parking position. Continued aircraft operations within the established framework must also be unaffected. Cockpit instruments, flight procedures, aircraft fuel tanks, pipes, valves and measuring and monitoring equipment, as well as the engines and auxiliary power units (APUs) of all aircraft types must operate without deviation from existing certifications. The fuel burn of any alternative aviation fuel must be similar to the JET A-1 fuel burn, allowing airlines to fly the same type of aircraft with the same tank capacity on the same routes at the same airspeeds to match existing flight patterns and to maintain slot times at congested airports for predictable aircraft operations without fuel-related delays or load restrictions.

From an environmental perspective, the climate-friendly jet fuel must meet all the requirements for acceptance as a certified sustainable aviation fuel that meets the emissions reduction requirements of the relevant governments. Finally, the price of SAF is of utmost importance to airlines, as their aircraft operating costs are mainly influenced by the price of kerosene, and any significant price change cannot be absorbed by the airline without leading to a significant subsequent increase in ticket prices, affecting flight bookings, booking classes and the potential risk of market distortion.

To add to the complexity of the fuel changeover, JET A-1 from petroleum and JET A-1 from SAF will have to coexist in the markets, logistics and airports for a period of more than 25 years, with product volumes of the same specification but with increased manufacturing costs, market prices and new supply chain elements. The logistics to the airport will become more complex and the environmental advantage of SAF can only be physically traced to the blended storage at an airport from where the jet fuel is delivered to all aircraft refuelling without separation of the JET A-1 SAF molecules from the fossil JET A-1 molecules. This requires a different way of tracking SAF volumes until blended with JET A-1, and then a chain of custody tracking of the remaining transport activities up to the aircraft wing fuelling valve for SAF blended jet fuel volumes.

From an SAF producer's perspective, the oil industry's seamless production pathway from crude oil exploration to distribution of refined products to the end user cannot be replicated. Different feedstock sources must be managed, including feedstock certification, and the production output – mainly synthetic green diesel and SAF – must be blended with its fossil counterpart before being physically transported to the distribution companies. Each participant along the supply chain needs to make a profit in relation to the intermediate value added they provide for the sale of the final product. The logistics chain is quite sophisticated and, in the absence of transfer pricing (which is used within large corporations to distribute the value added to a product), each participant has to charge its customer for the services and/or products it provides. In addition, the production processes for converting organic feedstocks into hydrocarbons are expensive and the size of most processing units does not allow the economies of large-scale production. As long as these conditions persist, the market price for SAF is at least twice the price of fossil Jet A-1, and for some pathways, three or four times the price of Jet Fuel.

From the buyer's perspective, a commodity product has an identical layout or specification to all competing products. In these circumstances, the purchase decision is directly related to price. SAF grades represent the new generation of aviation turbine fuels that have entered the jet fuel market, meeting the JET A-1 specification with the exception of lower density and no aromatics in most cases. SAF offers similar product characteristics to fossil JET A-1, but is two to four times more expensive, has no operational advantage over JET A-1 other than its improved environmental footprint and requires a multi-disciplinary approach to purchasing. In addition, the existing market model for jet fuel pricing, which uses spot market announcements for JET A-1 as the basis for pricing, does not work for SAF. The high price of SAF and the uncertainty as to how an airline should behave in a new market where established fuel purchasing procedures no longer apply has led airlines to refuse to purchase SAF until new market rules for SAF are established to provide a better basis for fuel purchasers to make decisions. In effect, SAF's entry into the market is taking airline fuel purchasers back to the 1980s, when contract negotiations were a face-to-face business and price changes had to be negotiated in writing or verbally during telephone calls or meetings between the parties throughout the term of a supply contract. Despite the fact that kerosene accounts for between 20% and 30% of an aircraft's operating costs, airlines don't have the staff to devote more time and effort to fuel purchasing. The dynamics of the market, combined with the complexity of the SAF, require excellent purchasing skills, extensive experience outside the commodity markets and the flexibility and openness to new types of supply agreements. In all likelihood, the redefinition of jet fuel purchasing will require an expansion of the job profile of jet fuel purchasers as a prerequisite for the successful establishment of SAF supply contracts.

SAF prices and the challenge of managing the uncertainties of the SAF market have resulted in some airlines conducting SAF demonstration flights to familiarise themselves with the new fuel type, but subsequently refusing to purchase significant volumes of SAF or to have such volumes blended by one of their fuel suppliers prior to delivery to their airports. While the first pathways for SAF were certified in 2009 (FT-SPK) and 2011 (HEFA), it has taken almost 10 years for aviation industry operators to commit to purchasing significant volumes of SAF to secure investment in new SAF production facilities and demonstrate their willingness to be pioneers in a new jet fuel market model.

To date, airlines have shown limited activity in mitigating global warming through SAF purchases, but have focused their environmental efforts on a range of fuel saving projects by reducing aircraft zero fuel weight (ZFW) and optimising flight procedures, in addition to purchasing new aircraft with improved fuel efficiency as a step towards reducing emissions. In addition, many airlines have integrated a voluntary carbon offsetting platform into their flight booking engine, where passengers are asked to offset the CO_2 emissions associated with their flight and booking class individually during the booking process, in addition to paying for their tickets. Amounts collected are used to fund third-party emission reduction projects or for voluntary purchases of SAF, which are later used to fuel subsequent aircraft.

From a manufacturer's point of view, marketing experts have yet to come up with a strategy to sell an unknown product that is invisible to the end user, costs more than twice as much as the current product, and whose only benefit is an invisible improvement in air quality through a significant reduction in greenhouse gas emissions. In a competitive environment where all airlines try to attract passengers by offering low fares in order to increase their revenues on selected routes with low demand, any reduction in the profitability of the route will lead to subsequent problems in maintaining such routes as part of the airline's network. SAF would have achieved greater market penetration if aircraft engines still produced visible exhaust plumes from the combustion of fossil fuels as they trailed the approach or take-off path around airports. Advances in modern engine design by aircraft engine manufacturers that mitigate this readily apparent evidence of environmental impact are delaying the acceptance of higher price levels for all types of SAF. With the unanimously accepted goal of decarbonising the transport sector, governments have been challenged to achieve the breakthrough of SAF through legislation that will create the necessary momentum to overcome the ongoing holding pattern in SAF deployment.

Since 2012, some governments have taken measures to introduce mandates (e.g. Norway and Sweden) or subsidies for the use of SAF in aviation, which will force or encourage fuel producers and airlines to gradually replace fossil JET A-1 with SAF rather than relying solely on offsetting CO_2 emissions by purchasing allowances as the cheapest option for emissions compensation. In parallel, the US government has established an SAF support mechanism based on voluntary SAF purchases, with the price premium covered by subsidies in the form of federal tax credits equal to the competitive disadvantage. In addition, the US government is undertaking various efforts to stimulate large investments in new domestic production facilities as part of the federal government's incentive programme to transition the US aviation sector to a climate-friendly aviation network over time (US Department of Energy, 2022).

Countries outside the US and the European Union are closely monitoring the various policy approaches before making a national decision. In the current environment, legislation in the United States and the European Union will set the regulatory framework and timeline for aviation's transition from fossil fuels to sustainable aviation fuels over the coming decades.

2 Crude Oil and Its Supply Chain from Well to Wing

Assuming that any market entry of an SAF product must be in line with globally accepted market conditions, a basic understanding of the jet fuel market is a prerequisite for the successful integration of SAF following the phase-out of fossil jet fuels. Furthermore, the oil market is the key determinant of the pricing and competitiveness of all types of jet fuel, including SAF. Given the importance of aviation, a shortage of jet fuel is unlikely. The use of SAF therefore will not be driven by anticipated supply constraints and is therefore only related to price competition as long as SAF-related regulations are not in place in either direction.

2.1 A BRIEF INTRODUCTION TO THE OIL MARKET

The development of industrial production in the 18th century marked the beginning of the carbon age. The availability of energy in unprecedented quantities, unlimited in time and constant in output was the cornerstone of industrial development. Coal was the first energy source of choice because of its unlimited availability, its ease of storage and the possibility of separating coal mining and industrial production in time and space. The railway industry began in 1825 with coal as the energy source for steam engines. Liquid fuels began to replace coal as the primary energy source for road transport from 1886 with the development of the first internal combustion engine and the first powered flight by the Wright brothers in 1903.

The so-called "oil age" began in the second half of the 18th century with oil drilling in Wietze (Lower Saxony, Germany) in 1855 and in Titusville (Pennsylvania, USA) in 1859. Crude oil consists of various hydrocarbon compound classes, mainly alkanes (paraffins), cycloalkanes (naphthenes) and aromatics. Crude oil is processed in refineries by distillation, which separates liquid fuel products by boiling temperature. The so-called middle distillates are used for power and heat generation (fuel oil) and in the transport sector as diesel for cars, trucks, railway engines, as kerosene for aircraft and helicopters and – as a heavy distillate – as heavy fuel oil for shipping.

Lighter products are used for petrol for cars and heavier products are used as heavy fuel oil for shipping. In 2019, global crude oil production will be 95 million barrels per day (Statista, 2023). The share of aviation turbine fuels of all oil products included in 2019 was reported to be 4.75 million barrels per day, or 5.0% of 2019 production (Statista, 2023).

Crude oil prices fluctuate based on supply, demand and speculation. The volatility of the oil market can be seen in the main grades of oil, WTI (West Texas Intermediate), UK Brent and the OPEC basket. Consumer demand for predictable prices for future deliveries has created the oil trading market. Oil traders offer fixed prices for specific volumes at fixed dates, called "futures" when commodity futures are agreed on exchanges (e.g. New York Stock Exchange – NYMEX), or "forwards" when agreed by off-market agreements. Because futures contracts are bought and sold several times before physical delivery of the oil product to the final buyer, trading in futures contracts exceeds physical delivery many times over. The main reason for these transactions is not price hedging, but price speculation with arbitrage profits. The result of this commercial activity is reflected in the spot market prices quoted on exchanges such as Rotterdam for the European oil market or New York for the US oil market.

Jet fuel pricing is decoupled from physical transactions because oil products reflect not only crude oil prices but also the competitive situation between different oil products. Jet fuel must therefore maintain its position in competition with road diesel and light heating oil. From an economic point of view, every refinery tries to maximise the profit it can make from the production of oil products. Modern refineries can vary the output of products derived from crude oil. If there is a high demand for road diesel, the price of road diesel will rise as a result of the demand situation. As a result, a refinery would increase its production of diesel and possibly reduce its production of jet fuel. The price differential between diesel and jet fuel will increase. Volume contracts prevent the refiner from shifting production to a higher diesel output, but he could increase the price of jet fuel to compensate for the restriction on diesel output and the spread-related loss of profit.

While these price movements determine the daily spot market prices for various oil products, the price volatility of a commodity product such as jet fuel is translated into supply contracts with monthly fluctuating prices based on the monthly average of all price notifications from the previous month. A countermeasure to such price movements is price hedging of dedicated volumes to protect the jet fuel budget from extraordinary price increases.

By definition, the oil market as such is characterised as an oligopoly. Only a few suppliers sell oil products to a polypolistic community of consumers. A polypolistic market contains many consumers without market power because they are dependent on supply and individually have no alternatives to influence prices or other options to substitute the required product with product alternatives. Typically, price elasticity in an oligopolistic market – such as the jet fuel market – is absolutely limited because airlines have no alternative to jet fuel regardless of price. An airline's overall demand for jet fuel will deteriorate as a result of the discontinuation of service on routes that suffer revenue losses due to fuel-related price surcharges to cover high fuel costs as part of total aircraft operating costs. Overall, airlines are caught up in oligopolistic fuel pricing and the associated reduction in demand through higher ticket prices. Oligopolistic behaviour in a commodity market can be said to occur when a supplier's pricing policy is determined not only by the price elasticity of demand but also by the pricing behaviour of competitors, who may be encouraged to follow similar price movements for the identical commodity product, unaffected by variations in demand.

2.2   CRUDE OIL UPSTREAM SUPPLY CHAIN

Crude oil has been used as a sealant for wooden ships since 12,000 BC. The Greeks used oil as a lubricant for axles and shafts and called it "naphtha". The name has survived to this day for light petroleum products used in the chemical industry. Large-scale US oil exploration began in Texas in 1870, following initial oil production in Pennsylvania, spurred on by John D. Rockefeller, founder of the Standard Oil Corporation, who began selling petroleum as lamp oil. At the same time, oil exploration began in 1871 in Baku, Russia (now Azerbaijan), by the Nobel brothers and the Rothschild banking family in Paris, who financed a railway line from Baku to the Black Sea for the rapid transport of Russian oil. In 1907, Marcus Samuel – a land developer hired by the Rothschild family – was sent to Sumatra (Indonesia) to explore for oil in Southeast Asia. Instead, he founded the Royal Dutch/Shell Group as an oil company and became a rival to the Nobel Corporation (Paeger, 2019).

Upstream oil and gas companies specialise in the exploration, drilling and production of crude oil and natural gas by operating drilling rigs and specialised equipment to capture crude oil streams pushed up by natural pressure, or by pumping crude oil to the surface at onshore locations and transporting crude oil by pipeline, rail or ship to refineries. Offshore exploration requires specialised vessels capable of drilling into the seabed. Crude oil is then transported by seagoing vessels. After "peak oil" in 2005, US companies began exploring and producing the so-called unconventional oil and gas fields in shale formations by injecting a mixture of water, sand and chemicals to force oil from these formations to the surface through hydraulic fracturing. This process is known as "fracking" and accounts for a significant proportion of domestic crude oil production in the US.

Oil fields have a limited life, which is related to the amount of oil found. Over time, the rate of production declines as the oil content of the field decreases. Conventional oil fields are economically recoverable to a maximum of 40% of the identified volume. As the oil fields are located in pressurised formations, eruptive oil recovery is the state of the art. However, by using special pumps, the total oil production from an oil field can be increased to 60% of the available volume, but at higher production costs. Upstream oil production ends at the refinery gate. The processing of crude oil is the first step in the downstream process.

2.3 OIL DOWNSTREAM SUPPLY CHAIN

Crude oil is distilled in refineries by boiling the cleaned crude oil under ambient pressure or vacuum into fractions. The fractionation process takes place in the so-called rectification columns and results in a defined mixture of substances that are converted into specified products by further separation steps. Modern refineries can vary their product yield by varying the distillation process. In addition, hydrocracking can be used to break down heavy oil residues into shorter carbon molecule chains, leading to the commercial use of previously unsaleable oil residues.

For this reason, petroleum products are referred to as "commodities" because their product characteristics are uniformly standardised by specifications to ensure that products from different producers do not differ in quality and can be mixed and stored together in the same tank (commingled storage). The only difference between commodities is the price at which they are sold. For reasons of efficiency, internal combustion engines are designed to run on a common fuel specification, making it necessary for these fuels to be available internationally.

Gaseous products from the distillation process, such as butane and propane, and light oil fractions, such as naphtha, are sold to the chemical industry as base chemicals.

The middle distillates group, which includes gasoline or petroleum, kerosene, diesel and light fuel oil, is the largest producer group in crude oil refining. Its products are mainly used in transport and as heating oil for domestic and industrial use. Heavy distillates are heavy fuel oils for heating and power plants that use water vapour to generate electricity through steam turbines and sell the residual heating water to a district heating network to heat apartment blocks, office and industrial buildings as well as to heat water for large greenhouse operators growing vegetables and flowers. Heavy distillates are also used to produce bunker oil for ship engines as an energy source for large ocean-going vessels, and bitumen as a base for asphalt. Due to the environmental damage caused by their combustion, high sulphur heavy fuel oils will need to be replaced in the foreseeable future as environmental concerns about their impact on climate change grow. The International Maritime Organisation's (IMO) LSF2020 regulation requires shipowners to take measures to reduce air pollution by either using low-sulphur fuels (LSF) from 2020 onwards, or installing an exhaust gas cleaning system (EGCS), or converting their engines for dual use of bunker oil and liquefied natural gas (LNG), with LNG to be used preferably in coastal areas (Hapag Lloyd, 2018).

Middle distillate fractions are the "cash cows" in a four-part portfolio matrix, as they are mature products with strong consumer demand, and the revenues generated also finance new investments in the "stars" of the four-part portfolio matrix, which represent the next generation of products that will replace the current "cash cows" over time (BCG, 2023).

From a commercial point of view, motor gasoline and road diesel are mass-market products distributed through company-owned service stations operated by franchise partners and delivered by transport companies operating at their own risk, but with equipment that must be operated in accordance with the safety procedures required by the relevant product seller. For the supply of heating oil, oil companies have entered into franchise agreements with distribution companies, which may act as resellers but are obliged to follow the instructions given by the oil company. Product sales agreements oblige the distributor to buy heating oil exclusively from the oil company, which supports its sales as a marketing partner. While private consumers have to pay their fuel bill directly at the petrol station where they fill their vehicles, large transport companies conclude direct contracts with oil companies after tendering for their annual fuel volumes by region or by country. In fact, road diesel and light heating oil are similar oil products, produced by an identical distillation process, but with different product characteristics in terms of additives and product colour for tax reasons.

Compared to automotive gasoline and road diesel, jet fuel production is a niche market, representing about 5%–6% of crude oil production, which will reach 92 million barrels per day in 2023. However, jet fuel is a premium product with special requirements for refining processes, storage conditions and quality assurance. Sales of Jet A-1 are handled by the head offices of the national subsidiaries of multinational oil companies or, in the case of national oil companies, by the sales department within the head office. Today, based on the total number of domestic and international aircraft movements at the 100 largest international airports, it can be assumed that approximately 60% of the world's jet fuel demand is supplied to these 100 airports.

When civil aviation took off in the 1960s, each oil company at European airports had its own tank farm and, in some cases, its own hydrant system. In the early 1980s, jet fuel suppliers began to optimise their jet fuel distribution. Airport depots were converted into joint ventures with all jet fuel suppliers as joint venture partners with equal voting rights. Off-airport supply was optimised by exchanging jet fuel volumes with the joint venture partners for different airports as a cost-saving collaboration that avoids unnecessary transport from different refineries to an airport. In addition, in the event of a jet fuel stock outage at one of the joint venture partners, the other partners normally borrow the required quantities from their stocks to avoid disruption to that oil company's airline customers. Borrowed quantities are only borrowed at the inventory accounting level of a depot. Once the oil company in question has arranged for additional supplies to be delivered to the airport, the incoming deliveries are used to return the borrowed quantities to their original owners and to settle all inventories in accordance with contractual obligations. Reasons for a zero-stock situation at an airport depot joint venture partner include unforeseen quality problems at a refinery leading to a suspension of supply from that refinery, excessive demand from an airline customer due to additional flights or a change in aircraft type at short notice, or technical problems or industrial action in the distribution network of the oil company concerned.

Airports in the US operate fuel depots either centrally or adjacent to a terminal complex. At many airports, fuel depots (and hydrant systems where applicable) are owned by airport authorities or airline joint ventures. In the case of an airline joint venture, shareholder voting rights at some depots are based on jet fuel throughput in a previous period rather than on equal shares of all airlines as shareholders.

US airlines that operate their own terminal at an airport typically provide aircraft ground handling services using either their own personnel or a specialised aircraft ground handling service provider. When airline personnel provide ground handling services, they are also trained to refuel an aircraft. The quantities of jet fuel delivered are recorded by the crew member performing the fuelling procedure and reported to the flight crew of the refuelling aircraft by electronic data transmission or by manual transmission of the fuel slip to the ramp agent or cockpit crew. Quantities taken from a hydrant system shall be reported electronically to the depot for inventory adjustments. Refuelling by bowser follows the same procedure as hydrant refuelling, but the quantities delivered may be reported to the depot with a delay until the bowser returns to the depot to refuel with jet fuel at the depot fuelling station. The tax exemption for jet fuel requires consistent accounting for the inventory. Jet fuel used for engine test runs, maintenance flights and pilot training is not tax exempt and so the depot operator must frequently report fuel deliveries for non-commercial purposes to the tax authorities.

Jet fuel suppliers wishing to sell their product at an airport must apply for a sales permit from the airport authority and agree to the airport's rules of use. Permits to sell jet fuel are usually granted free of charge and for an unlimited period. Based on such a sales permit, a new entrant needs access to the depot and hydrant system, as well as staff and equipment for aircraft refuelling.

Under the Airport Usage Regulations, jet fuel suppliers can apply for throughput at the depot and hydrant system without having to become a joint venture partner in the fixed facilities. In addition, the jet fuel supplier may enter into a service agreement with one of the aircraft fuelling services at the airport for the fuelling of the aircraft of its (future) airline customers. Prior to commencing physical deliveries to the airport, the new supplier must register with the tax or customs authorities to sell tax-exempt jet fuel. Airport depots must provide access to third-party throughputters on non-discriminatory terms. Throughputters and depot shareholders must agree to the terms and conditions of an airport jet fuel depot. Such an agreement is known as the "Consortium Indemnification Agreement" or, in honour of the editor of this agreement, the "Tarbox Agreement". Companies using the depot for their sales activities must provide evidence of aircraft liability insurance coverage of up to 2 billion US dollars. This coverage can be obtained from brokers or by demonstrating a financial reserve of USD 2 billion that can be made available at short notice to provide immediate cover in the event of an insurance claim relating to the quality of jet fuel or an accident on the airfield that may be related to the hydrant system or the operation of trucks on the tarmac. The rationale behind this arrangement is to provide immediate indemnification to the party who was the seller of the jet fuel to the aircraft involved, regardless of whether an agent or employee of the seller was the person or party who caused the damage while acting on behalf of the seller. The Indemnity Agreement Consortium requires claims to be settled promptly. Thereafter, the cause of the incident will be investigated and the responsible party identified will be required to reimburse the seller of jet fuel – or its insurance company, as the case may be – who was obligated to settle the case under the terms of the Indemnity Agreement Consortium. This provision has a significant impact on SAF's deliveries to airport fuel depots.

Similar rules apply to US airport fuel depots. Based on airline "ex-refinery" purchases, jet fuel delivered to the airport fuel depot must be insured by the airline that owns the fuel. The airline must obtain and maintain insurance coverage for liabilities arising from deterioration of fuel quality during blending as well as for incidents and accidents involving jet fuel on the tarmac or in flight. The majority of insurance claims are related to the operation of vehicles on the tarmac. In many cases, a fuel dispenser or bowser collides with another vehicle, or the bowser collides with the engine cowling or airframe. Fortunately, these types of damage can be identified before the aircraft takes off and, in most cases, personal injury – if any – is limited to outpatient treatment of ground staff and does not affect the safety of passengers and crew.

2.4   THE ROLE OF STOCK MARKETS AND TRADING OF JET FUEL

Contractual reference to stock exchange notifications for traded jet fuel has become the main tool to reduce the workload during the execution period of a jet fuel supply contract. Several publishers report daily market prices for different products. It should be noted that JET A-1 is not traded daily on all exchanges, but the publishers provide prices for each trading day regardless of the number of transactions. On days when there is no physical transaction of jet fuel, the publishers call sellers and buyers to ask them their current price expectations. The information gathered from these calls is condensed into a virtual price, which is intended to reflect the actual market price if a transaction had taken place. Acceptance of such a process requires a high level of trust in the objectivity and independence of the publisher from all parties involved.

Stock market prices allow sellers and buyers to enter into longer-term supply contracts without being at a competitive disadvantage compared to short-term contracts, which may start at a more attractive price level, depending on the interest of the party wishing to avoid the negative effects of a long-term commitment. From an economic point of view, the production base of any commodity should be sold under longer-term contracts in order to secure the economic viability of the production facility for the seller and to provide a financial incentive to the customer who is also willing to place its base volume with one supplier without any detriment.

Airline fuel purchasing departments that are familiar with the handling and storage of jet fuel may not only enter into "ex-refinery" purchase agreements but may also purchase the so-called "cargoes" on the spot market. Depending on the market situation, JET A-1 refined product is shipped in a tanker, i.e. from the Arabian Gulf to the North Atlantic Ocean without a designated port of destination. The fuel on board is marketed during the voyage on the basis of the best price offer received for the volume carried. At the end of the voyage, the tanker may be directed to a designated US or European port, or it may be anchored offshore for a period of time. The shipping company is responsible for arranging berthing at the designated port and all customs procedures. The buyer must arrange for pipeline transportation and storage capacity at a JET A-1 storage facility near the port or, in the best case, direct on-carriage to the destination.

As long as market prices reflect a realistic relationship between supply and demand, the communicated prices provide a fair contractual basis that does not favour any party. However, the global communications network reports within minutes on major events such as natural disasters, unexpected refinery closures, transport route disruptions and military actions that affect the supply chain for oil products. In fact, any news item that could potentially affect the supply situation for oil products in general or jet fuel in particular is immediately reflected in commodity price increases, and even if such an event is limited in scope and any potential supply restriction can be covered by spare capacity at other refineries or by diverting ocean-going vessels to other routes, which may only result in a delay in arrival of a few days, such information leads to a jump in commodity prices within hours. As long as the potential threat to supply persists and is not removed by countermeasures or evidence of miscalculation, stock markets will remain in disarray. Consequently, such unwarranted price movements work to the advantage of sellers, as commodity prices will reflect the incident in the following month's commodity prices. The increase in natural disasters, conflicts, terrorist attacks and pirate activity in the Red Sea, the Gulf of Aden and around islands in Indonesia, Malaysia and the Philippines that affect the transport of oil products, as well as the fear of war triggered by isolated military events, have contributed to price levels for crude oil that are unrelated to the availability of the product, the impact of capacity constraints or unexpected consumer demand.

Airlines pay an involuntary premium for being contractually obliged to accept all price movements caused by third parties but reflected in the jet fuel exchange price. IATA airlines have made several attempts to replace the automatic price adjustment with formula pricing, but with little or no success. While in the 1980s the concept of formula pricing was considered to be a perfect tool for the efficient execution of jet fuel contracts, the airline industry has paid a high price for the convenience of easy contract execution. On the other hand, when kerosene prices determine up to 30% of the operating costs of an airline fleet, these costs remain as an industry phenomenon beyond the influence and decision-making of airline boards. When oil prices are high, airlines collectively introduce fuel surcharges on tickets that passengers have to pay – nowadays, following various court decisions, as an integral part of the overall ticket price in line with consumer protection laws. The airlines' unpleasant fuel cost situation has led to apathetic behaviour: As long as high fuel prices affect the entire airline industry, such periods do not lead to competitive disadvantages for an individual airline. This behaviour is a logical consequence of the sector's total dependence on oil market movements, oil production and refining infrastructure, and the huge investment that would be required if airlines were to start producing jet fuel "in-house". Without the necessary background and expertise in jet fuel production, the opportunity cost of such a manoeuvre would remain prohibitively high. In this respect, SAF, with its completely different production pathways compared to crude oil refining, could become a game changer for aviation. New technologies, new feedstocks and new players will provide a unique opportunity for airlines to start vertically integrating their jet fuel supply chain as long as the new SAF market is open to new supply models. Investment opportunities in SAF production facilities may offer higher returns than additional investment in new aircraft fleets, based on the existing profitability of major airlines with corresponding fuel demand. As a result of fuel price escalation and volatility, the cost of jet fuel is a critical success factor for any airline.

3 The Jet Fuel Market – JET A-1 and SAF

Jet fuel is a commodity product produced in refineries to international or national specifications. In addition to JET A-1, the corresponding military jet fuel is JP-8, which has a slightly different specification with the addition of corrosion inhibitors and anti-icing additives, but is fully interchangeable with JET A-1. JP-8 is primarily used by NATO air forces.

JET A is identical to JET A-1 with the exception of the freezing point. While JET A-1 has a guaranteed freezing point of −47°C/−52.6°F, JET A has a freezing point of −40°C/−40°F. JET A was mainly produced and used in the US, while other countries preferred to use JET A-1, which is the dominant type of jet fuel today. The huge volume of civil jet fuel consumption in the US, which uses JET A as the standard jet fuel type, must have its economic background in a price differentiation between JET A and JET A-1/JP-8. JET A-1 is now widely used throughout the US for the refuelling of civil and military aircraft. The extended range of modern aircraft types allows flight durations in excess of 12 hours in a cold atmosphere, with the aircraft gradually climbing to higher flight levels to conserve fuel once the aircraft operating weight has been reduced by the amount of fuel already burned. Operating the aircraft in extremely cold temperature conditions will have a cooling effect on the fuel tank temperature, and before reaching the freezing point of the fuel, the pilots will be forced to descend to lower flight levels where the ambient temperature is less cold, thus avoiding the risk of engine failure due to rapid freezing of the fuel in the wing tanks.

In this respect, the ambient temperature at the departure airport becomes an issue. Jet fuel is stored in large tanks that are exposed to ambient temperatures through the metal surface of the double-walled tank shell. Depending on the settling time of the fuel in the tank, the stored jet fuel can be allowed to cool down over a period of 24–72 hours prior to refuelling the aircraft. If the fuel temperature is low at the time of refuelling during the winter season, the fuel temperature will continue to decrease during take-off as the aircraft passes through flight levels where the temperature conditions are already lower than the fuel temperature in the aircraft. Tank heaters can slow down the temperature drop, but cannot stop the cooling effect. In this respect, SAF offers a lower freezing point than fossil JET A-1 and improves fuel-related flight procedures.

In principle, an airport's fuel demand is largely determined by the expected scheduled operations of all the airlines serving the airport. Airline fuel departments receive fuel uplift forecasts from their flight operations department based on a calculation of the flight routes departing from an airport, the aircraft models operating those flights and the frequency of flights on a route. The airline's weight and balance computer system calculates not only the amount of fuel required for an individual

DOI: 10.1201/9781003440109-3

flight but also uses these figures for statistical purposes and to forecast weekly or monthly fuel requirements by combining the flight plan with historical fuel uplift statistics. These figures reflect a standard refuelling operation in accordance with regulatory requirements for additional fuel for holding, diversion, missed approach and contingency. The airlines' route profitability analysis includes, among other cost elements, the cost of fuel based on the prices paid during the month. At airports with extremely high fuel prices, "tankering" of jet fuel is a common practice to reduce fuel costs. In proportion to the price difference, the aircraft will take more fuel than necessary on the inbound flight in order to avoid or even limit the refuelling of the aircraft for the outbound flight at the airport where suppliers charge high fuel prices. The airline's weight and balance system calculates not only the additional amount of fuel to be loaded in accordance with the aircraft's weight limits, but also the additional fuel burned for such flights in relation to the additional amount of fuel carried and the additional fuel cost attributable to the additional amount burned. The additional cost of tankering on the inbound flight must be offset by the cost savings attributable to the reduced amount of fuel to be lifted on the outbound flight. Tankering can create a cost spiral as the airport's fuel infrastructure will handle less fuel. Such a reduction in volume will increase the unit cost of kerosene handling and have a knock-on effect on prices, leading to even more inbound routes where tankering provides a cost advantage to the operating airlines. The same threat is posed by varying levels of SAF mandates at international airports. An increase in the price level of jet fuel at the airport caused by an SAF blending mandate with higher unit prices than the price of fossil jet fuel will lead to increased tankering activity by airlines on their inbound flights to airports affected by high SAF mandate levels. Such a response will result in an increase in fuel consumption in terms of tankering volumes, while at the same time reducing the uplift and the total amount of SAF required to meet the mandate in the first place, as intended by the government in charge. Both effects are detrimental to the objective of reducing greenhouse gas emissions. At least the EU Commission is considering a new requirement for all airlines to refuel at least 90% of the fuel calculated as regular uplift at an airport as a legal provision to prevent airlines from tankering. However, it is questionable how such a rule could be monitored and how misbehaviour could be sanctioned, since the amount of fuel on board is at the discretion of the captain under the law of the country that has issued his airline's Air Operator Certificate (AOC) for his airline. At the very least, governments outside the EU Member States will not accept any provision that restricts their sovereign rights when their airlines fly into EU Member State airports.

3.1 THE GLOBAL NATURE OF JET FUEL AND ITS IMPACT ON MARKET DESIGN

Historically, the oil industry has been able to distil crude oil into user-specific products – called "specialities" – that have revolutionised the transport sector. The vast distances within the US, the high infrastructure costs of highways and railways, and the long travel times associated with long distances by rail and road laid the foundations for US air travel. At the same time, US oil companies grew as oil fields were developed and consumer demand for gasoline, diesel and jet fuel increased. The integrated "well to wheel" supply chain contributed to an optimised transportation fuel supply structure that is the backbone of US domestic jet fuel distribution today (Figure 3.1).

The US domestic and international aviation market is by far the largest jet fuel market in the world, with total consumption of 1,743,380 bbl/day (80.9 million tons/year) in 2019. Based on global jet fuel consumption of approximately 290 million tons in 2019, the US share was 27%, followed by China with 36.8 million tons and a share of 12.2%. Simultaneously, jet fuel costs reflect an important cost factor of aircraft operating cost without any option for airlines to substitute this cost element (Figure 3.2).

Based on an expected market growth of 2% per year, the market dynamics lead to a unique situation: (1) Assuming that suppliers are interested in maintaining their market share in a growing jet fuel market, their strategy requires a constant change in refinery output by expanding jet fuel production or adding more production facilities to their portfolio of jet fuel supply sources, or agreeing with a competitor on product borrowing and product balancing by supplying jet fuel to their customers in other regions. (2) The supplier may alternatively increase its production of jet fuel while reducing its production of other middle distillates. (3) A supplier limited to its current annual volume of jet fuel may maintain its volume but may be willing to accept a reduction in its market share if such a solution does not result in an absolute reduction in profits. (4) In line with limited refining capacity, the responsible supplier may choose to purchase additional jet fuel volumes from competitors at marginal profit

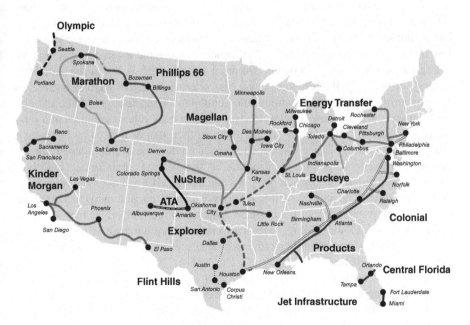

FIGURE 3.1 US pipeline system – major pipelines carrying jet fuel.

Source: Airlines for America (2022) (prepared by author).

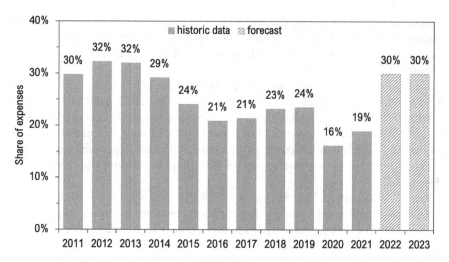

FIGURE 3.2 Fuel cost share of aircraft operating cost 2011–2021 with forecast 2022 and 2023.

Data source: IATA, S&P Global Platts; Statista 2023 (prepared by author).

to maintain its market presence, or (5) as a future option, purchase SAF from SAF producers to fill its supply gap and maintain its market share by blending SAF with its JET A-1 in line with government incentives or mandates.

In a continuously growing market, the intensity of competition is different from a saturated market where a loss of volume requires immediate action to regain market share. Typically, saturation in commodity markets leads to discounting of a competitor's customers as a countermeasure to induce customers to compensate for lost volume by switching suppliers. The sequence of volume losses and discounting increases the likelihood of a downward price spiral, which is not in the suppliers' interest. From a distribution perspective, the price elasticity of jet fuel is close to zero, as airlines need jet fuel without the possibility of substitution. While the airlines naturally seek the lowest price, the battle to win back volumes reduces margins, while at the airport affected by such a price conflict, the airlines enjoy temporary relief on their jet fuel budget. The need for action on the sellers' side is less dramatic as long as the sellers can meet their internal jet fuel volume commitments built into the company's annual production plan. As long as the seller can meet his company's profit expectations, he may be willing to accept a small reduction in his market share as long as the profitability of his business remains within budget. He may consider volume stability at lower selling prices with a reduced profit margin as a second-best option, which may also result in a reduced annual reward for his business success as a personal consequence.

When a new supplier joins the existing pool of suppliers at an airport, competition for customers and volumes will increase as the new competitor will take some customers and their jet fuel volumes away from the existing suppliers.

The introduction of SAF as an SBC for JET A-1 may not have an immediate impact on the local jet fuel market, as the SAF supplier will initially be required to have its SAF volume blended with JET A-1 prior to delivery to the airport.

However, airlines may choose to enter into independent SAF supply agreements and have their purchased SAF volumes blended at the facilities of an oil company that is the current supplier to an airline. Consequently, such airline behaviour will have an impact on jet fuel supply volumes but will not jeopardise the current market price level for JET A-1 at the airports concerned. Once airframe and engine OEMs have developed a process that allows SAF to be used as a pure jet fuel without blending requirements, such a change in ASTM JET A-1 and SAF certification will have a direct impact on the volume distribution at airports. The price differential between fossil JET A-1 and SAF remains an issue, but as long as governments support the voluntary purchase of SAF through subsidies that cover the price premium between the fossil JET A-1 price level and the SAF market price, such additional volumes can cover the market volume growth. Countries with legislation requiring an SAF mandate for jet fuel have decided that the so-called distributor (i.e. the party obliged to pay taxes on the sale of products – i.e. Value Added Tax, VAT) is responsible for achieving the greenhouse gas reduction by gradually replacing fossil energy sources with renewable energy sources in order to comply with the maximum permitted emission level. Therefore, the legal obligation to mitigate climate change is not on an airline, i.e. the polluter through its engine exhaust, but on the seller of jet fuel who sells JET A-1 to an airline.

As is common practice for oil companies selling JET A-1 on a "free into-plane" contract, the origin of the crude oil and the name and location of the refinery that produced the JET A-1 are not disclosed to customers for several reasons: (1) The supplier may vary the source of the JET A-1, by shipping from more than one refinery. (2) The supplier may have a product exchange agreement with another oil company and the JET A-1 is delivered to the airport depot by a competitor but sold under the supplier's brand name. (3) JET A-1 is imported from abroad and the name of the refinery must be kept confidential. Consequently, the seller of the blend is not obliged to disclose the name of the producer of the SAF if he is obliged to blend JET A-1 with SAF-SBC regardless of whether the SAF is produced in-house or purchased from a third party.

Conclusion: Under current conditions, the jet fuel market is considered to be an airport-centric business in countries where jet fuel is mainly supplied by oil companies offering "free into-plane" supply contracts. In contrast, the supply jet fuel supply in the US is seen as a domestic issue with different conditions at hub airports and major airports due to the nature of the supply and the status of the US domestic airlines, which play an active role in the jet fuel market.

3.2   PRICE FORMATION IN THE PROCESS CHAIN FOR JET A-1

While "costs" refer to cost categories such as internal production, services, logistics provision and expenses, the "price" of a product or service is an expression of its value. The so-called "market price" originated in ancient times, when traders and buyers would meet at certain times in the centre of a city in a place suitable for conventions. Such events were called "markets" and the place was called "market place". The "quoted price" for a product was the result of such events, when buyers and sellers physically met and negotiated prices. With a number of traders and sellers offering identical goods in the same place, consumers had the choice of finding the best deal for their needs. Such trading activity led to a balance between quality and quantity and the associated prices for such goods. In addition to quality and quantity as the main determinants of price, availability also became an issue. Scarcity rent reflects the price premium that consumers are willing to pay for the immediate availability of a product. Excess supply, on the other hand, leads to price erosion if the product is not storable or if the storage costs are likely to exceed the expected sales profit. Maximising profit is the main task of all economic activity. Conversely, it requires minimising losses.

In the long term, the cost of producing crude oil and equivalent products is expected to rise. For economic reasons, oil companies have initially developed and exploited the easiest fields with the lowest production costs. Depending on the size of the field, the production life of crude oil is between 20 and 50 years; fracking reduces this to between 2 and 5 years at most production sites. The economic recovery is about 60% of the existing reservoir volume. Replacing depleted oil fields requires the development of new production sites, whose exploration and production costs are typically higher than those of the depleted fields. The WEF's production cost forecast for 2020 assumes daily production of around 80 million barrels per day from OPEC, other crude oil producers and the states of the former Soviet Union (FSU), with production costs ranging from $15 to $30 per barrel. Production above this level, up to 100 million barrels per day, requires production costs of around $40–60/bbl. Further development of oil fields, oil sands and oil shale, as well as the production of residual volumes from abandoned oil fields, is in the range of around USD 80/bbl (World Economic Forum, 2011). As the cost of oil production shifts to fields with higher production costs, while global demand for oil continues to grow, the price floor for crude oil will also rise. At the same time, oil companies tend to anticipate a medium-term decline in demand for petrol and diesel for passenger cars, due to the shift from internal combustion engines to electric drives. As a result, the economics of replacing ageing refineries are more challenging and new refineries will only be built if they can ensure a sufficient return on invested capital. As a result, the utilisation of existing refineries will continue to rise and global spare processing capacity will decline, leading to higher diesel and jet fuel prices.

The main determinant for kerosene is the development of the crude oil price. Irrespective of how the markets react to short-term developments, the oil price forecast is based on the economic environment of the global economy. The extraction of raw materials, industrial production and the transport sector all require correlated amounts of energy. In addition, the demand for electricity and heat is also largely met by the use of oil products. The second determinant of kerosene prices is the oil price policy of groups such as OPEC or non-OPEC countries with large oil production capacities, such as the countries of the former Soviet Union (FSU), Norway or the US. While OPEC countries derive most of their GDP from the export of crude oil, GDP in industrialised countries is generated by the production of goods and services. This creates a conflict of economic interests: Oil-exporting countries seek to maximise the selling price of crude oil. Developed countries, on the other hand, need to keep the price of oil at a level that does not undermine the competitiveness of their industrial production. The US, in particular, has steadily increased its crude oil production in recent years and will be the world's largest producer in 2022 with

11.8 million barrels per day, followed by Russia with 10.6 million barrels per day and Saudi Arabia with 10.3 million barrels per day.

Price formation in commodity markets depends on several factors: In oligopolistic markets with identical products, the price is determined not only by supply and demand but also by the exclusionary behaviour of competitors. A supplier can increase the selling price of its product as long as customers are willing to pay that price. If competitors also increase their prices, their market shares will remain unchanged. If competitors do not follow the price change for their products, the supplier risks losing customers to the cheaper competitors. Conversely, competitors must also accept price reductions by the supplier or risk losing their own customer base. The jet fuel market also operates in this tension. New entrants that reduce or close any supply gap in petroleum-based products by substituting products expand the supply side and thus the price responsiveness of the suppliers in the oligopoly! Oligopolistic market behaviour exists whenever the pricing policy of a firm depends not only on the (price-elastic) response of the demand side but also on the price response of competitors, who can lower their product prices to maintain their market share even without additional sales expectations (Figure 3.3).

The "equilibrium price" in an oligopoly (pA in Figure 3.3) is therefore the maximum achievable price "A" at which no supplier in the oligopoly is inclined to change its own price because it can expect its competitors to reduce their prices in the same way and thus all suppliers would only reduce their profits if their market shares remained unchanged. Conversely, an isolated price increase (above "A" and the associated price level pA) will not lead to profits if the other suppliers do not follow suit in the short term. In such a case, the unilateral price increase would shift demand to the competitors and increase their cumulative sales through the loss of sales by the initial price-increasing supplier. Therefore, the common interest of the suppliers

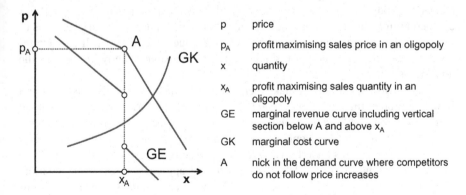

p	price
pA	profit maximising sales price in an oligopoly
x	quantity
xA	profit maximising sales quantity in an oligopoly
GE	marginal revenue curve including vertical section below A and above xA
GK	marginal cost curve
A	nick in the demand curve where competitors do not follow price increases

FIGURE 3.3 Nicked demand curve in oligopoly situations [Woehe (1981): General introduction into business administration (prepared by author)]. Explanation: p, price; pA, profit maximising sales price in an oligopoly; X, quantity; xA, profit maximizing sales quantity in an oligopoly; GE, marginal revenue curve including vertical section below A and above xA; GK, marginal cost curve; A, nick in the demand curve where competitors do not follow price increases.

necessarily lies in a monopoly-like pricing behaviour, where each supplier behaves in such a way that the profit situation of the entire industry is maximised for a given distribution of market shares, thus avoiding a price war.

From the airlines' point of view, the supply oligopoly leads to an influence on the operating costs of their aircraft fleet that is largely beyond the influence of a professional purchasing organisation, since all jet fuel sellers have a similar objective: to maximise prices while avoiding market share shifts. Taking into account the common practice of airlines tendering their 12-month requirements on a regional basis, a similar behaviour has been observed on the demand side.

Based on their respective demand forecasts, airlines tender their demand for each airport in a region at the same time in order to implement individual bargaining strategies. If an airline were to tender its jet fuel requirements ahead of time, jet fuel suppliers would not offer aggressively calculated prices because there would still be sufficient options in the subsequent tender wave to sell the planned sales volume to other customers at better prices. If the airline waits too long to tender, suppliers may have already sold their planned volumes to other customers and therefore have no incentive to offer a better price to the latecomer. If all suppliers have already contracted their volumes, the price may even rise, putting pressure on an airline's purchasing department to sign a contract.

For airlines, the optimal negotiation strategy is to start price negotiations for the airports in a region at the right time to identify, through market monitoring and negotiation, those suppliers who are under volume pressure and therefore willing to offer discounts or other price reductions to get a deal done quickly. Price negotiations reveal the core competence of any buyer. He needs to know or at least be able to assess the suppliers in a region, the competitors who supply there and the prevailing market situation. The purchasing strategy described here is based on the dynamics of the air transport market. Routes are opened, expanded, reduced or abandoned depending on the intensity of demand and the competitive situation. Aircraft, as a means of production, are used flexibly by airlines. If there are too many airlines competing in a market and the route is unprofitable, i.e. at least the variable costs of operating the route are not covered by ticket revenues, the airline will either offer fewer frequencies per week, use smaller aircraft or abandon the route and use the aircraft elsewhere.

Against this background, it makes sense to tender fuel supply contracts only for a limited period of time in order to be able to negotiate better prices when fuel demand increases. Since the "free into-plane" contract model does not guarantee a volume purchase, it is also in the suppliers' interest not to enter into longer-term contracts where demand remains uncertain and refinery capacity must first be reserved.

Jet fuel purchasing in the US takes a different approach. In domestic aviation, much of the demand is stable because, in the absence of a dense rail network, air travel is the mass transport mode for the domestic population and also the best alternative in terms of time and price relative to the distance travelled. In this respect, the conditions for a stable route network for American airlines are incomparably better than in other countries. As a result, passenger demand on individual routes is also subject to constant demand, with comparatively small fluctuations due to holiday travel. Demand is also subject to cyclical fluctuations, but these affect all domestic routes equally. In this respect, demand for jet fuel on domestic routes is less volatile than in Europe, and the price elasticity of passengers is also lower, since for distances above 250 miles/400 km there is no alternative means of transport and the car is the only alternative. Under these conditions, US airlines can contract for fixed volumes of jet fuel because they are needed to serve a geographically limited market, and the buying airline can count on a steady flow of traffic to the airports in a region, even over a longer period of time. Even if they shift demand between airports, excess volumes can be diverted to other airports or, failing that, sold to other airlines. If reselling results in a lower loss (relative to the previous purchase price) than the additional transport costs of diverting the excess volume to other airports in the airline's network, this may be the preferred means of minimising losses. This is also the case for any resale at less than the previous purchase price. Depending on the characteristics of the aviation market, there is therefore no universal recipe for how suppliers and buyers of jet fuel can find each other. Given the increasing demand for jet fuel and the unit cost of the product, buyers and sellers move large volumes of product and millions of dollars. Because suppliers and buyers do not meet every day in a real marketplace, all market participants must rely on the reliability and trustworthiness of their counterparts.

Another element of oil price formation is speculation on futures exchanges. It is expected that speculation in the futures market is also based on the expected economic development of the world economy. However, there is no clear positive correlation, as the development of the oil market in 2021 and 2022 shows. As a result of the Covid19 lockdown, industrial production and the transport sector were massively reduced, which was accompanied by a decline in oil production and low oil prices due to temporary oversupply. After the end of the Covid19 restrictions, the temporary shortage of oil supply was caused by the longer start-up time for oil production and processing, which could not keep up with the speed of the increase in transport and production growth. With the start of the war in Ukraine in 2022, the rise in the price of crude oil continued, as uncertainty about the availability of sufficient quantities of oil became the determining factor for the price. Despite the ongoing Ukrainian crisis, the oil price fell in the second half of 2022 as analysts saw the risk of a global recession, which in turn put pressure on the oil price (Lynx, 2023).

It is in the nature of the complex oil market that price movements cannot be accurately predicted due to the various influencing factors. By concluding a supply contract based on a formula price, the price of the product can be fixed at the time the contract is concluded, but the price is then subject to the monthly fluctuations of the stock market. Airlines plan their production output for a given period in terms of accumulated passenger or freight kilometres based on revenue per kilometre assumptions. In addition, airlines plan a cost budget in which the cost of jet fuel plays a significant role. The airlines' primary interest is to keep the assumed fuel price within the planned budget and to buy as cheaply as possible. Given the volatility of the market, this is no easy task. Therefore, financial products from the capital market are used to compensate for price fluctuations. This is known as fuel price hedging. The following two examples show how price hedging works: (1) The airline buys a collar for a defined quantity of jet fuel, i.e. a price band with an upper and lower price limit for a defined period of time. If the kerosene price is within the collar, the airline pays only the contracted kerosene price. If the price exceeds the upper limit of the collar, the airline pays its supplier the contract price, but is reimbursed by the collar provider for the difference between the upper collar price and the contract price paid. Conversely, if the contract price falls below the lower limit of the collar, the airline pays the contract price to its supplier but must pay the difference between the lower collar price and the contract price to the counterparty of the financial product. (2) Alternatively, the airline can hedge its jet fuel budget using a cap. In this case, the airline agrees an upper price limit with the broker for a defined quantity of jet fuel over a defined period. If the price of the fuel exceeds the cap price, the supplier reimburses the airline for the difference between the contract price and the cap price. In return, the broker charges a premium equal to the amount of jet fuel he has hedged. The cap fee covers the difference between the expected average market price and the cap price (set as the upper limit that the airline wishes to hedge). The closer a cap price is to the expected market price, the higher the premium payment, as there is a greater likelihood that the cap provider will be required to perform.

The difference between the expected market price and the cap price is crucial in determining the benefit of a cap. As the gap between the market price and the cap price above it increases, the probability of a payout and the amount of the payout decrease. As a result, the amount to be paid when a cap is taken out will also decrease. If a cap is used to hedge a fuel budget, it is necessary to check at what fuel price level the budget will be exceeded. A distinction must also be made between short-term price spikes and a permanently higher price level. The amount of the contract price for the cap must be included in the benefit assessment.

In addition to the jet fuel price hedging options described above, the financial market offers a wide range of price hedging transactions. Depending on the contract model, these contracts involve processing fees and corresponding payments to the financial service provider. They can be used to smooth out price fluctuations within a budget period. However, long-term participation in the price development of the jet fuel market is unavoidable.

Conclusion: The formula pricing model for JET A-1 sales "free into-plane" allows airlines with a small purchasing team to meet their global demand for JET A-1. However, airlines pay a high price for this convenience. Buffering price spikes through congruent hedging has a short-term dampening effect on prices, but cannot neutralise price speculation. The emerging SAF market offers a unique opportunity to move to a new pricing model. If airlines do not seriously consider this option, the SAF market will also end up with a quoted formula price for supply contracts.

3.3 PRICE FORMATION IN THE PROCESS CHAIN FOR SAF

The ASTM-approved production routes for SAF are divided into (1) processes where the kerosene has a density below the JET A-1 minimum density and contains no aromatics, and (2) processes where the kerosene meets the JET A-1 density requirement due to the process technology and also contains the aromatics required by the specification.

SAF grades according to (1) do not fully meet the specification requirements for JET A-1 and are therefore only approved as a synthetic blending component

(SAF-SBC). They must be blended with fossil JET A-1. SAF grades according to (2) meet the JET A-1 specification and can theoretically replace fossil JET A-1 as a neat fuel without the need to blend. However, due to a lack of practical experience with JET A-1 compliant SAF, this production route was initially only approved as a blending component. Airframe and engine manufacturers are working on an improved jet fuel specification that will allow SAF to be used directly as a modified JET A-1 in the future, eliminating the need for blending with fossil JET A-1. No date has been set for the use of SAF as a neat product. Given the need to rapidly replace fossil JET A-1 with SAF, it is likely that the current restrictions on use will be lifted in the coming years.

The following discussion of SAF pricing is based on the current approval requirement that SAF may only be used as a blending component and must always be blended with JET A-1 prior to delivery to an airport depot.

At the beginning of 2023, there is still no consistent and industry-wide accepted pricing model for SAF. Given the heterogeneity of SAF production processes, the first step is to consider pricing based on an SAF producer's cost of production. Each SAF production path is based on process-specific feedstocks, some of which need to be pre-treated and then stored at the production site. In the production process, synthetic fuels, synthetic naphtha and waxes are obtained from these raw materials through several process steps. (1) If the raw materials are purchased from third parties, a delivery price must be paid for them, including transport to the processing site, which is included as a cost element in the cost accounting and thus in the price calculation. In the case of integrated production, the raw materials are obtained from own production and are then calculated with a transfer price. (2) Subsequently, the costs of warehousing the raw materials, processing them and storing the synthetic fuels produced until they are delivered to a customer are all considered to be production costs and are also included in cost accounting. Therefore, the cost of conversion consists of the cost of feedstocks and the cost of processing. (3) The extended cost of conversion also includes the cost of transportation to the airport jet fuel depot or, alternatively, the cost of transportation to the blending depot where SAF is blended with JET A-1, the cost of marketing and distribution, the producer's administrative costs, and (4) in the case of a "free into-plane" delivery contract, the cost of interim storage at the depot and the cost of the "into plane" fuelling service for aircraft refuelling. For simplicity, the extended SAF production costs are considered according to (3). These cost elements correspond to an "ex-refinery" sale, as offered for petroleum products in the US, plus the downstream costs to be added. A mark-up for profit and risk must also be calculated on the extended production costs according to (3) resulting in the final sales prices (4). In addition, it is necessary to check whether the procurement costs of the raw materials and the production variables "electricity", "gas", "water" and "intermediate products" are subject to price fluctuations during the period under consideration. The range of variation of these variables has to be determined and an additional cost to be derived from it has to be estimated.

This method starts by determining the production price at the beginning of a delivery period. Depending on the volatility of the production variables and the duration of a supply contract, a SAF pricing model is then needed that congruently reflects the volatility of the input materials and thus ensures consistent cost recovery and profitability over time.

On the basis of pre-existing purchase contracts with (1) airlines, (2) oil companies or (3) marketers, the SAF producer makes its synthetic kerosene available for transport "ex works" (EXW) in accordance with Incoterms 2023 or delivers it "free on board" (FOB) or "cost, insurance, freight" (CIF), which includes the transport costs to the agreed delivery point as part of the delivery. The delivery point need not be a vessel for loading, but may be a suitable tank farm that meets the storage requirements for JET A-1.

In the US model, where an airline purchases an agreed fixed quantity of SAF, the airline arranges transport and blending with JET A-1. At the time of blending, both kerosene grades are owned by the airline.

In the EU model of variable "free into-plane" deliveries, the oil company providing the JET A-1 purchases the SAF either in its own name or in agreement with an airline as an agent and trustee of the airline. If both types of synthetic kerosene are owned by the oil company, ownership of the blend is clear and the oil company sells the blend successively when refuelling its customers' aircraft, provided it is a mandatory blend under mandate. If an airline also purchases SAF on a voluntary basis, these must also be physically blended with fossil JET A-1 before they can be stored at an airport depot.

If an oil company acquires ownership of a batch of SAF and subsequently blends it with JET A-1, the oil company will sell the blend to its customers under the respective contract model for fossil JET A-1. As long as the blends are in the single digit percentage range, the price distortion caused by the allocation to the fossil jet fuel price is not significant. For higher blends up to the current 50% blend wall, the innovative SAF product would be sold at oil market conditions and thus exposed to speculative prices unrelated to the product.

The illustration of the SAF price impact in Tables 3.1 and 3.2 shows the price per cubic metre (264 US gallons) of a JET A-1 blend in USD at various blending levels for HEFA and PtL SAF grades. A 30% HEFA-SAF blend already increases the price per cubic metre of jet fuel by 60% compared to pure fossil JET A-1.

Under the assumptions of Table 3.1, the 2022 effective fuel expenses based on a 30% HEFA blend of the following legacy carriers would have to be increased:

- United Airlines: from 13,113 million USD to 20,980 million USD,
- Delta Airlines: from 11,500 million USD to 18,400 million USD,
- Air France/KLM: from 7,241 million EUR to 11,586 million EUR and
- Lufthansa: from 3,792 million EUR to 6,067 million EUR.
  (United Airlines, Delta Airlines, Air France/KLM, Lufthansa, 2023)

From the oil companies' point of view, the continuation of the spot market-based formula price is a simple solution for price updating. As SAF volumes increase in the market, a synonymous formula price model for SAF based on the average quotations of the previous month may be established as a follow-on solution in the future. From the airlines' point of view, such a model would miss the unique opportunity to abolish the speculation-driven spot market model and establish a cost-based pricing formula. The premise that the price reflects the value of a product can only be implemented to a limited extent for SAF, as the value added is intangible and the

TABLE 3.1
Illustration of SAF Price Impact: Jet Fuel Blended with HEFA up to Maximum Blend Ratio of 50% Based on Theoretical Product Prices

Price [US$/cbm]		Blending Percentage [%]		Price [US$/cbm]	Price Increase [%]
Fossil jet fuel	SAF (HEFA)	Fossil jet fuel	SAF (HEFA)	JET A-1 blend	JET A-1 blend
800	2,400	100	0	800	0
800	2,400	98	2	832	4
800	2,400	95	5	880	10
800	2,400	90	10	960	20
800	2,400	80	20	1,120	40
800	2,400	70	30	1,280	60
800	2,400	50	50	1,600	100

Source: Prepared by author.

TABLE 3.2
Illustration of SAF Price Impact: Jet Fuel Blended with PtL up to Maximum Blend Ratio of 50% Based on Theoretical Product Prices

Price [US$/cbm]		Blending Percentage [%]		Price [US$/cbm]	Price Increase [%]
Fossil jet fuel	SAF (PtL)	Fossil jet fuel	SAF (PtL)	JET A-1 blend	JET A-1 blend
800	4,000	100	0	800	0
800	4,000	98	2	864	8
800	4,000	95	5	960	20
800	4,000	90	10	1,120	40
800	4,000	80	20	1,440	80
800	4,000	70	30	1,760	120
800	4,000	50	50	2,400	200

Source: Prepared by author.

value of SAF is measured independently of the JET A-1 market price. The lack of measurability of the correlation between product value and product price means that the value of SAF is mathematically negative.

As long as the jet fuel market is dominated by the volume share of fossil JET A-1, the pricing of SAF is limited to the sale of SAF production and can be freely agreed between producer and buyer under these conditions. With the removal of the blending requirement by ASTM and the permitted use of additive enhanced SAF as a JET A-1 equivalent, those SAF producers will initially become direct sellers whose production processes already meet the requirements of the JET A-1 specification ASTM D1655. Subsequently, the other SAF production routes will also be granted 100% use as JET A-1 substitutes. Under this condition, all SAF producers will be able to sell their synthetic kerosene directly to airlines without any intermediaries.

From a scientific point of view, the direct entry of new SAF producers into the aviation market has to be considered under two conditions: (1) differences in production costs between JET A-1 and SAF due to different raw materials, feedstock sourcing and processing technologies, and (2) differences in production costs between different SAF production processes.

In the entry phase of an SAF producer, the extended cost of production represents the long-term price floor. As each SAF production path uses different feedstocks and has a different ratio of feedstock to production costs, the individual products differ only in their production costs, but not in the composition of the fuel, as all processes must meet the specification requirements. Given the high price level for SAF relative to the fossil JET A-1 price, there are two strategic options for SAF producers: (1) An attempt is made to keep the SAF price as low as possible through volume scaling in order to minimise the difference to the JET A-1 market price in the long term. This strategy is based on the expectation that lower prices will increase voluntary demand for SAF outside the mandates and lead to further expansion of production. (2) In view of the rather large spread between the JET A-1 market price and the SAF market price, an attempt will be made to establish an independent SAF price level in the aviation sector that is not dependent on the JET A-1 price. The advantage of this solution is that significantly more investment capital will be available for the construction of new production facilities and the provision of the required quantities of raw materials. As the approved SAF production processes, with one exception, are still at TRL 5–7, the further scaling up of the plants to industrial production (TRL 8–9) still represents an uncertainty factor, which has to be reflected in a higher return on investment. The accelerated construction of new plants, which is desired due to climate protection requirements, will take place all the more quickly the better the return expectations of investors can be met.

Over the past 12 years since the introduction of the first two SAF production pathways, airlines have refrained from actively participating in the product maturation of the SAF production pathways and, within their means, from acquiring SAF. The market demand for SAF has so far been artificially created mainly by government subsidy programmes – especially in the US – while in Europe, individual SAF projects have been specifically promoted without generating a self-sustaining demand. The main argument of the airlines against purchasing SAF volumes was and is the purchase price. Since it will not be possible in the foreseeable future to produce SAFs at the manufacturing cost of JET A-1 and thus reach the JET A-1 price level, the unit cost degression for large systems will not induce airlines to purchase SAFs voluntarily, unless they receive government incentives to compensate for the price premium for SAFs or are forced to do so by their customers or by mandates from their governments. Instead, airlines will offer passengers the opportunity to voluntarily offset their share of CO_2 emissions by adding a surcharge to the ticket price when purchasing a ticket for the selected route, thus giving passengers the feeling that they are travelling in a carbon-neutral manner. In addition to the established fare categories, "green fares" are now available as a booking option, which include a CO_2 offsetting surcharge as an integral part of the ticket price, depending on the class of service. According to the airlines, the revenue from CO_2 offsetting is used to purchase SAF in batches from time to time, which are then flown at a later date.

This airline policy is an additional challenge for the SAF market, which makes it difficult to penetrate the market with this innovative product. New approaches have to be developed to motivate airlines to actively participate in the SAF market. Pricing plays an important role in the SAF market as it is the key driver for market penetration.

Assuming that the ASTM SAF blending regulations will be lifted in the foreseeable future, the role of SAF producers as equal market participants in the supply of aviation jet fuel must also be considered. Compared to the incumbent jet fuel suppliers, who have more than 50 years of experience in marketing jet fuel, SAF suppliers, as new entrants to the market, have to compete with established competitors.

According to market theory, the market equilibrium of the incumbent suppliers in a petroleum product oligopoly is disturbed by the entry of new suppliers, unless the new SAF suppliers adapt "ab initio" to the existing market practices, which is unrealistic in view of the significantly higher production costs for SAF. In this respect, it can be assumed that the start of industrial production of SAF will initially have little or no impact on the JET A-1 price, as the volumes put into circulation are small and the higher price level will have no impact on the market equilibrium of the fossil jet fuel market.

In principle, equilibrium parity can be assumed if new market entrants (1) limit their production volumes to the increase in demand and thus do not change the existing volume relationships between the established market players, (2) price themselves at the market equilibrium price – if they are able to do so – in order to avoid unwanted price reactions from the established suppliers, and (3) market SAF and its significantly higher price level as a substitute product "with additional environmental benefits for the demand side": These benefits are the CO_2 emission credits that can be obtained for SAF or the government subsidies available to compensate for the SAF price premium.

The direct marketing of SAF through bilateral contracts between SAF producers and airlines has already started in the US market, as airlines and SAF suppliers can jointly agree on the blending procedure with JET A or JET A-1. In all other countries, direct marketing of SAF to airlines may follow as soon as ASTM lifts the blending requirement with fossil jet fuel. However, the 100% SAF revisions to ASTM D7566 that the jet fuel subcommittee is already working on will still require blending, but with a different numbering of the Annexes affected i.e. due to changes in additives required. This will be necessary to meet the compositional and property requirements of Jet A or JET A-1 as the case may be.

As SAF products enter the market, the question of different production costs of SAF in relation to the raw materials used and the conversion costs becomes more important.

The market equilibrium conditions of the established JET A-1 market apply equally to a new SAF market: Assuming that new SAF suppliers are interested in stable market conditions in which they can build market share without price wars, the necessary condition for "soft" entry into a new market is a sufficiently high SAF equilibrium price that covers the production costs of alternative fuels plus a satisfactory return on capital. As long as the unit cost of producing alternative fuels is below the SAF equilibrium price, all affected SAF producers can offer at the SAF equilibrium price without competition and thus achieve different but positive profit margins. This requires an undersupplied SAF market, where all production can be sold without competition. As the market matures and the supply gap closes, the SAF market will change when demand is fully satisfied and price competition develops. However, if the production costs of an SAF production process are above the SAF equilibrium price, economically viable production can only be represented if the unit costs of these SAF production routes are artificially lowered by direct subsidies to enable market participation. This requires a special interest to start such a production process. In principle, it would also be possible to artificially raise the SAF equilibrium price. However, this would be an arbitrary intervention in the existing competition regime and could be challenged under competition law as an illegal concerted price cartel. The break-up of the Standard Oil Company in the US in 1911 could serve as a precedent (Figure 3.4).

Ceteris paribus, the SAF market as a commodity market will also become an oligopoly in the long run. In the initial phase, independent producers will compete with each other, but they will have to choose one of the already approved production paths as their production technology. As a result, associations will be formed to represent common interests and, as demand increases in all aerospace markets, mergers or acquisitions will take place as small SAF suppliers will find it difficult to establish a global presence.

Theoretically, the question of price reduction in the oligopoly to prevent further SAF producers from entering the market must be considered, especially since the established oil companies can also decide to develop their own SAF production with high capital investment and use their financial strength to take market share from new market entrants. Price dumping is prohibited under competition law, but the combination of integrated fossil jet fuel and SAF marketing certainly offers opportunities

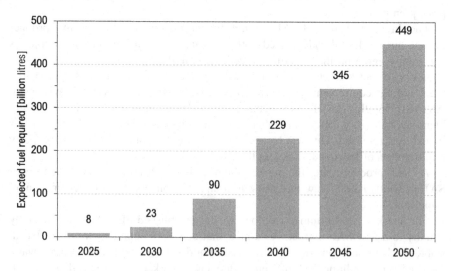

FIGURE 3.4 Expected SAF required to achieve Net Zero in the aviation industry in billion litres.

Data source: IATA-fly net zero; Statista 2023 (prepared by author).

to create supply constellations that are not objectionable under competition law and that neat SAF producers will not be able to offer to their customers. The "anticipatory" price reduction follows the principle of damage limitation: the falling SAF price (in established markets) shifts the break-even point for new suppliers, which may prevent or at least make it more difficult for them to enter the market. In the analysis of alternatives, the loss of profits resulting from lower sales revenues at constant production volumes must be compared with the loss of profits resulting from the decline in production volumes due to a loss of market share as a result of market entry by new competitors.

For the production of SAF, three key economic questions are therefore at the forefront of the investment decision: (1) In what period of time will SAF theoretically be able to achieve a complete substitution of fossil JET A-1? In this context, an assumption must be made as to the market share in the jet fuel market at which an SAF equilibrium price changes to an SAF price competition and how the respective SAF production paths will then compete with each other. (2) How will the price of crude oil develop globally (demand-oriented) and in relation to the production volume (supply-oriented), i.e. how will the equilibrium price for the existing products develop in the future projection? The level of the crude oil price will have a decisive influence on the duration of the transition phase from fossil JET A-1 to SAF. (3) Can SAF be marketed coherently with fossil JET A-1 as a complementary product in the existing JET A-1 market model, and what barriers to entry need to be overcome to achieve this?

These three questions describe the tension between fossil JET A-1 and SAF. Since the demand for fuels is largely inelastic to price, passing on cost increases in the price does not lead to a significant reduction in demand. SAF is therefore produced for the following reasons:

1. Expected returns on biomass and/or SAF production margins under the assumption of long-term rising crude oil prices.
2. Expected returns on the assumption of an actual or expected blending mandate that will allow substitute products to enter the JET A-1 market, including the costs of establishing their own sales and distribution channels.
3. The increase in supply of fossil fuel substitutes under arbitrage considerations up to the profit-maximising level of production – in the theoretical model up to marginal revenue equals marginal cost, where at the revenue maximum, the average return on investment is inevitably lower than the return on investment at the profit-maximising level of production.

This is not due to rising production costs, but to the speculative reflection of excess demand, where demanders accept rising prices and pass them on to their consumers in order to gain access to a scarce product. This happens irrespective of whether there is an actual physical shortage or only an anticipated one. Behind this price curve, alternative fuels can enter the market if their production cost plus a reasonable return on investment equals the long-term guaranteed scarcity price of the conventional oil-based fuel. Expectations about the use of alternative fuels to offset excess demand determine the future price curve, and thus, the investment in research, development, feedstock allocation, processing and, ultimately, the economic output of alternative fuels.

The determination of the scarcity of resources in terms of quantity and the derivation of the resulting price evolution for crude oil and downstream crude oil products is a matter of concern for academia and industry alike, as it is necessary to determine the right investment timing for the establishment of alternative fuel production, which on the one hand offers competitive advantages over other suppliers (as a first mover) and at the same time enables profitable production (usually attributed to the early followers).

Declining supply in the face of steady or even rising demand leads, over time, to a shortage and thus to further price increases, not because of rising production costs, but as a speculative reflection of excess demand, where demanders accept rising prices and pass them on to their consumers in order to gain access to a scarce product. This happens whether there is an actual physical shortage or only an anticipated one. Behind this price curve, alternative fuels can enter the market if their production cost plus a minimum return on investment equals the long-term guaranteed scarcity price of the conventional oil-based fuel. Expectations about the use of alternative fuels to meet excess demand determine the path of the future price curve and thus the investment in research, development, feedstock allocation, processing and, ultimately, the economic output of alternative fuels.

For the market introduction of alternative fuels, it is therefore necessary to forecast the future development of the price of crude oil as accurately as possible, to adopt legislative blending quotas or to introduce alternative legislative subsidy schemes. If the market introduction is too early, the minimum rate of return cannot be achieved or, if the price is prohibitive, demand will not respond to the alternative supply. It is also important to identify the combination of raw materials and processing methods that is available and technologically feasible. Such a combination must guarantee a reliable and predictable return on investment in relation to the price of fossil fuels and possible alternative production methods for SAF, at least over the payback period of the investment.

Conclusion: Pricing a new product in a new market creates uncertainty for manufacturers and airlines as buyers. If the market develops differently from the assumptions made by the parties to a multi-year contract, there is a risk of misallocation of resources. At least in the start-up phase of a new business model, the lack of mutual benefits can be mitigated by close cooperation between the parties. A cost-plus contract model with disclosure of production costs and/or airline participation in the capital of the production company can mitigate the risks for both parties.

3.4 SAF CONTRACT DESIGN

The different SAF production pathways require a new type of contract for the purchase of SAF, regardless of whether SAF is purchased as SAF-SBC or for direct use as a JET A-1 equivalent. Currently, most SAF production pathways are based on biomass-based feedstocks. The requirements for the feedstocks used in the production of SAF differ between the sustainability requirements of the US government and those of the EU Member States, which are required to transpose EU directives into national law. Compliance with the relevant regulations in the distributor's country is therefore critical for the subsequent eligibility of the produced SAF for subsidies or indirect tax credits in the US, as well as for crediting in the European Emissions Trading Scheme.From a supplier's perspective, it is recommended that the following issues be considered prior to entering into an SAF supply agreement with an airline:

1. The selection of raw materials suitable for processing in the conversion plant is of great importance in the SAF supply agreement. Depending on the technology selected, raw material costs can account for up to 70% of SAF's total manufacturing costs. A low cost feedstock will therefore have a direct positive impact on the budget calculation. However, it should be analysed how volatile such raw material prices are and what price increase can be expected in a worst case scenario within the supply period of the sales contract.

2. Media interest in SAFs has focused on the commodity aspects of biodiversity in the context of monoculture biomass cultivation, the food versus fuel controversy surrounding the potential displacement of food crops and the issues of land use change and indirect land use change caused by the displacement of food crops on arable land by energy crops. Further to an independently certified Life Cycle Analysis (LCA) of the feedstocks procured, it is recommended that the feedstocks used should be specified as precisely as possible. In particular, if the physical identity between the raw material used and the respective production batch – which is normally not relevant for commodities – is in question, the raw materials should be congruently allocated to the production batch on the basis of a mass balance and this allocation should be documented accordingly.

3. The recycling of residues and wastes as a source of raw materials is becoming increasingly important as an alternative to conventional biomass cultivation, especially as the use of these materials avoids issues of biodiversity, food competition and land use. This is underpinned by the expectation that these recovered materials cannot be used for anything else. However, in some cases, multiple options are being developed for the reuse of these waste materials. As a result, a new market for waste and residues will emerge, characterised by price competition, once the waste material has become a product for which there is a market and price competition begins. Those willing to pay the highest price will gain access to the materials. As prices for waste and residues increase, the impact on SAF will be reflected in corresponding price changes, which may lead to the introduction of a price escalation formula as part of an SAF supply agreement.

4. In the case of blending arrangements with fossil JET A-1, the maximum blend ratio that can be achieved will be determined by the aromatics content, density and distillation characteristics of the fossil JET A-1 used for blending. The blending process should be contractually regulated and, with respect to liability coverage, explicitly included in the Intoplane Liability Insurance of the seller of the blend.

5. Sellers of SAF should be aware that oil companies may be willing to voluntarily compensate an airline with fossil JET A-1 in the event of a shortfall in the supply of SAF. The ability and willingness of the oil industry to act as a fallback option for SAF suppliers will be reduced as SAF gains market share. It is expected that the oil industry will introduce a charge for holding a reserve at refineries to cover storage costs and interest on the capital invested in the reserve.
6. In view of the SAF production chain with different process stages and different actors, the provisions for dealing with cases of force majeure also need to be amended.
7. A SAF purchase contract will initially follow the US model of a fixed quantity with partial deliveries at fixed dates. It will be necessary to determine whether the sale is to be made "ex refinery", with all further steps being the responsibility of the buyer, or whether the SAF producer is to deliver to a blending facility (fob delivery) in order to complete the blending with fossil JET A-1.
8. A special case is the sale of SAF to an airline with the involvement of an oil company: An airline concludes a purchase contract with an SAF producer for a defined quantity at a defined price. The virtual transfer of ownership takes place when the aircraft is refuelled. The SAF producer first sells the agreed quantity to the oil company at the price agreed with the airline. In agreement with the oil company, the SAF price also includes the cost of transport to the airport, storage at the airport fuel depot and the services of the refuelling service. The oil company thus owns the SAF and mixes it with fossil JET A-1. The mixture is then delivered to the airport fuel depot and the in-flight refuelling service virtually fills the airline's aircraft with the specified blend. On the basis of the quantities delivered at each fuelling, the oil company shall charge the airline the price of fossil JET A-1 (according to the formula price), the agreed price for the SAF quantity delivered as SBC and a handling fee for carrying out the blending at its premises.
9. All other provisions of a standard jet fuel supply contract also apply to an SAF purchase contract.

From an airline's perspective, the following recommendations should be considered when entering into an SAF supply agreement:

1. Strategic corporate positioning should precede the decision to purchase SAF. For this reason, an airline should primarily clarify its position with regard to the five dimensions of societal role, environmental issues, social issues, corporate image and economic framework conditions. The results of this positioning have a direct impact on the design of the purchase agreement for SAF. In addition to the commercial content of the contract, such an agreement must also contain specific terms and conditions for SAF's procurement in order to legally safeguard its own sustainability concept and to oblige the respective suppliers to comply with specified standards.

2. In the early commercialisation and market penetration phase, there are usually no industry standards that can be assumed to be generally known and recognised by means of a standardised framework agreement. In order to avoid disruptions in delivery and performance later on, buyers and sellers of SAF should therefore agree on clear rules and avoid any room for interpretation through broadly defined regulations. Particularly in the context of a heterogeneous supplier structure in the supply chain and raw material production in developing and emerging countries without transparent economic structures and without state control of cultivation and processing procedures, the buyer of the product must anchor retrograde quality assurance in the purchase agreement.

3. The requirements for the raw materials used in the production of SAF differ between the sustainability requirements of the US administration and those of the EU member states, which have to transpose EU directives into national law. Therefore, compliance with the respective regulations in the country of the distributor is decisive for the subsequent crediting of the produced SAF for the granting of subsidies or indirect tax credits (RINs) in the US as well as for crediting in the European Emissions Trading Scheme (EU-ETS).

4. For deliveries of neat SAF or SAF-SBC, the raw materials used, their geographical origin and the transport route to the processing unit shall be clearly verified for each production batch. This shall include evidence of the sustainability certification of the whole supply chain and a life cycle analysis, which may be based on the specific values of the supply chain or use default values. In addition, the supplier should guarantee that the production has been carried out in accordance with the UN Compact, excluding child labour, and that the applicable occupational health and safety regulations for the workers involved are observed and monitored. In the case of raw material cultivation in developing countries, it should be demonstrated that the cultivation of energy crops does not have a negative impact on regional water management and food supply.

5. With regard to raw materials, the airline must decide on the contractual exclusion of certain feedstocks, i.e. food products. Other reasons for exclusion may be justified by land use. Palm oil is mainly grown in Indonesia and Malaysia on former rainforest land. In this context, it is recommended to consider the acceptance of RSPO certified palm oil from sustainable cultivation. It is also recommended that, in the interests of accountability, the airline purchasing SAF should obtain an overview of the origin of the raw materials used. This includes evidence of sustainability certification accredited in the EU and a life cycle analysis showing a reduction in greenhouse gas emissions of at least 70% for inclusion in the EU-ETS.

6. If the SAF supply agreement is intended in particular to offset emissions in air freight in order to meet the corresponding requirements of the air freight forwarders for emission-neutral freight transport, the SAF supplier must prove his contractually guaranteed ability to deliver. In the event of a delay in delivery by SAF, he must be obliged to compensate for the missing SAF quantity by procuring a replacement.

If the quantity can only be compensated with fossil JET A-1, the supplier must provide the airline with alternative emission reduction certificates free of charge. During the ramp-up period, airlines will be able to compensate for any SAF shortfall with fossil JET A-1. The willingness of oil companies to compensate for the failure of an SAF supplier to deliver its product on time and at the agreed quality will diminish as the market penetration of SAF grades increases. In this respect, it is advisable to contractually regulate the consequences of a prolonged interruption of supply and, especially in the case of SAF start-ups, to contractually regulate the frequent reporting of key financial figures.

7. All other provisions of a standard jet fuel supply agreement also apply to an SAF purchase agreement.

The above recommendations show that the use of SAF as an environmentally friendly fuel requires much more elaborate supply agreements than the standardised industry supply agreements with a defined scope of provisions, where only the (estimated) quantity, the price formula and the location differential for the duration of the contract in the broadest sense need to be inserted. In terms of content, aviation is confronted with a more comprehensive responsibility for the entire process chain, from the production of the raw material to the delivery of the product to the aircraft, even if the production and processing of the raw material is not part of the contractual content of a fuel supply agreement. Ultimately, the media and public opinion will turn to the airline that purchases SAF if incidents or violations occur in the SAF supply chain. Comparable situations in the global textile and footwear industries show that respect for human rights, occupational health, safety and working conditions will always be attributed to the buyer of SAF who is in a position to exert the strongest influence on the manufacturer. Ultimately, even comprehensive contracts are no guarantee of adequate protection against reputational damage. However, they will pave the way for industry-wide accepted standards and procedures within the SAF supply chain.

Conclusion: The public perception of the success or failure of an SAF production is particularly evident in the case of the airlines as the end users of the product: damage to the image in the SAF process chain leads to immediate consumer reactions and thus generates indirect consequential costs, which not only relate to the elimination of the causes of damage in the process chain but also burden the business model of the airline concerned through loss of revenue and loss of market share. It is therefore in the interest of buyer and seller to avoid such damage through a very precise contract design and contractually agreed inspection rights of the buyer along the process chain. In 1995, for example, Shell's intention to dump the wrecked Brent Spar oil storage platform in the North Atlantic led to protests by environmentalists and a boycott of Shell petrol stations. By selling its excess production to competitors, Shell limited the economic damage. However, the damage to the company's image was reflected in the migration of car owners to other brands of petrol stations, even after it became clear that the planned dumping of the platform was environmentally safe.

3.5 RELIABILITY REQUIREMENTS FOR FEEDSTOCK SUPPLIERS AND SAF PRODUCERS

Reliability, in the sense of the legal obligation to fulfil a SAF supply contract, consists of several elements that interlock like the cogs in a classic clockwork mechanism.

From the airline's point of view, it is important that the fuel supplier fulfils its obligation to supply jet fuel reliably, punctually and without restriction on the basis of the contracts concluded. This is particularly true for "free into-plane" supply contracts, which cover all services up to the fuel inlet valve on the aircraft wing, including short-term additional jet fuel requirements if the number of departures and the size of the aircraft change during the term of the contract. The Jet Fuel Supplier is therefore obliged to scale and plan its production sufficiently to be able to deliver continuously and to have sufficient quantities of JET A-1 in stock at the airport fuel depot at all times to supply its customers. In the event of unforeseen interruptions in production or supply, the supplier shall maintain contingency plans to ensure the supply of an airport fuel depot in the event of irregularities or emergencies. A supplier's reliability also includes regular quality monitoring of its jet fuel throughout the supply chain to the point of aircraft refuelling. In the event of a supply disruption, regardless of its origin, the supplier is obliged to inform its customers immediately so that flights can be rescheduled and technical stopovers for aircraft refuelling can be planned. If the supply disruption affects the operation of the airport fuel depot, suppliers are obliged to arrange for the publication of a NOTAM (Notice to Airmen) via the relevant aviation authorities, informing all airlines of the operational disruption at the affected airport.

With regard to the supply of SAF, the reliability requirement covers the entire process chain from raw material production to delivery of the manufactured SAF. In the case of biomass-based feedstocks, crop yields can vary from year to year and may even fail completely. The SAF producer should therefore spread the raw material risk by distributing the purchase quantities among several raw material producers in different regions in order to secure its own raw material supply. In addition, the SAF producer is responsible for ensuring that the raw materials are sourced only from certified cultivation and that the certificates, which are limited in time, are renewed in good time by means of re-certification. In view of the new plant technologies involved in SAF production, longer downtimes for repairs and maintenance work are to be expected at the start of operations. Organic feedstocks can only be stored for a limited period of time. This requires storage facilities that allow uninterrupted plant operation.

There are also reliability requirements from the point of view of jet fuel producers to their customers, the airlines: The most accurate possible forecast of fuel requirements for a contract period is an important prerequisite for the jet fuel supplier's production planning. The scheduled performance of all published flights, as well as the frequency of arrival delays and flight cancellations, has a direct impact on the capacity of the fuelling services, which have to refuel each arriving aircraft within a short time window, and thus on the cost of aircraft refuelling. It is also very important to meet agreed payment dates for jet fuel invoices and thus demonstrate financial reliability. SAF start-ups, in particular, depend on timely receipt of payments to ensure their own liquidity and to meet the payment deadlines of their raw material suppliers.

Conclusion: The goal of an uninterrupted supply of suitable feedstocks for SAF production, as well as a smoothly operating plant technology and transport chain, requires not only the integrity of the people involved but also the willingness of each company involved as an integral part of a process chain to provide precisely those services in terms of time, quality and quantity that are necessary for the overall success of the process chain.

3.6   AVOIDANCE OF FEEDSTOCK SHORTAGES

SAF can only produce to the extent that the necessary raw materials are available. Any shortfall in the supply of contractually agreed raw materials leads to an immediate – and usually irreversible – reduction in the production output of a production facility, resulting in a financial loss and a loss of customer confidence. The possible strategies to avoid such production losses due to feedstock shortages inevitably have an impact on manufacturing costs: (1) By increasing raw material stocks, a time buffer can be created in which the supplier delivers shortfalls or replacements are procured from the market at short notice. This requires additional investment in storage infrastructure, increases storage operating costs and results in a higher capital commitment for the SAF producer for the stored quantity. The storage capacity of biogenic feedstocks is limited without conservation measures. On the producer side, storage of oleaginous crops leads to a reduction in oil content over time. On the processing side, there is a risk of quality deterioration due to overstorage. (2) Spreading the required feedstock quantities among several producers, with optional additional quantities to be called up as needed, reduces the risk of shortages. However, the SAF producer will not receive the best available purchase price for smaller quantities. (3) The SAF production capacity can be designed to be larger than what is required in normal operation. The standby capacity can be used to produce at least a backlog quantity, which can partially compensate for the loss of revenue. Higher investment and maintenance costs are incurred. (4) The SAF storage capacity on the supply side can be increased. The additional storage space creates an SAF buffer. The additional tank space also requires a larger investment, leads to higher operating costs and the capital commitment increases by the raw material and manufacturing costs. (5) As an option for controlled feedstock supply, at least the base load of feedstock requirements could become part of an integrated "farm-to-fly" concept that includes investment in agriculture as well as investment in the production facility in a holding structure or as a joint venture of two companies with reciprocal participation in the respective partner company. In this model, the variability in production output should be covered by feedstock supplies from the market and the base load will be part of a "closed-shop model" between the partners.

Shortages in raw material supply are not only due to meteorological conditions, pests and diseases, but also due to lack of labour and equipment, seed quality and farm management deficiencies, as well as irregularities in certification resulting in non-compliance. SAF produced from uncertified raw materials is not eligible for CO_2 credits in emissions trading. The feedstock requirements differ between the US Renewable Fuels Standard (RFS 2) for emission credits under the Environmental Protection Agency (EPA) in the US and the crediting regulations in the EU under the

EU Renewable Energy Directive (EU-RED II). The qualification requirements are so different that a feedstock certified in the US will not be eligible in the European Union. Conversely, an EU-certified feedstock could theoretically be accepted in the US, but the feedstock would also have to meet specific conditions. In this respect, the limited availability of feedstock in the European SAF market cannot be compensated by feedstock supplies from the US, unless the feedstock in the US is certified according to EU regulations, which is unlikely. As a result, any lack of domestic feedstock availability in EU Member States will have to be compensated by imports. Due to the existing crediting requirements, European SAF production will face feedstock certification issues that will have a negative impact on SAF production capacity over time, which can only be addressed by switching to non-biogenic feedstocks. These feedstocks are called Renewable Fuels of Non-Biological Origin (RFNBO), such as electricity-based SAF produced from hydrogen and carbon.

In the US, the feedstock for SAF is mainly sourced from the domestic agricultural sector, which allows for independence from raw material imports and at the same time secures jobs in the agricultural sector. In addition, the production of SAF takes place in manufacturing facilities built and operated in the US. This creates a domestic value chain for SAF, which in principle corresponds to the beginning of American crude oil production in the 19th century, but with different companies at the individual stages of the value chain.

While the average population density in the US is 37 inhabitants per square kilometre, the comparable figure for the EU is an average of 118 inhabitants per square kilometre, which is three times higher than the average population density in the US. The agricultural area in the US is 186 million hectares, with an average farm size of 170 hectares. In the EU Member States, the agricultural area is 172 million hectares, but the average farm size is only 14.3 hectares. As a result, biomass availability is greater in the US than in the EU countries. In 2022, 338 million people will live in the US, compared to 446 million in the EU-27. As a result, food crops require more arable land in the EU than in the US, and arable land for energy crops is limited. In addition, EU member states have a long tradition of importing raw materials and energy from other countries rather than producing them domestically. As a result, there are no efforts in the EU to achieve complete energy autarky through domestic production of all forms of energy needed. Therefore, biogenic raw materials for energy production also have to be imported from abroad. With the 2009 Renewable Energy Directive (EU-RED) and its 2018 amendment (EU-RED II), the EU Commission has set strict requirements for the sustainability and quality of energy feedstocks and their processing. At the same time, the use of biomass in the transport sector has been restricted in order to accelerate the transition to electric mobility in road transport. For SAF production in Europe, this results in significantly more complex raw material procurement routes with lower profit margins than in the US.

In addition to the use of agricultural biomass, waste and residues are also used as raw material sources. The use of cascades makes it possible to reuse materials that have already been used or to treat waste without creating additional carbon sources. The so-called "circular economy" relies on carbon cycles to minimise the further accumulation of CO_2 in the atmosphere.

The usable energy content of biogenic raw materials varies. Vegetable oils have a high energy content for a low volume and high specific weight. Conversely, wood residues and grasses have a low energy content and a high volume. For a given energy content, their mass ratio is significantly worse than that of vegetable oils. Against this background, it is understandable that the first generation of SAF production pathways focused on these two feedstocks, which were readily available in the start-up phase of SAF production in 2010.

The processing of vegetable oils and animal residues is very similar to the processing of crude oils in a distillation process. However, vegetable oils and animal fats have to be pre-treated in order to obtain comparable distillation cuts into light, medium and heavy distillation products in a boiling distillation. The ease of obtaining vegetable oil by pressing or extraction, the good storage and transport properties and the technologically well-controlled conversion to synthetic fuels have made oils and fats the world leader in the selection of raw materials for SAF production. With regard to the calculation of production costs as described above, vegetable oils and fats are leaders in terms of market prices because, even before their use for energy purposes, there was an exchange-listed market for vegetable oils that included rapeseed and soya for animal feed in addition to palm oil and sunflower oil as edible oils. Under these conditions, the emerging market for synthetic diesel and synthetic paraffin did not develop as a fully integrated production chain, but was limited to the operation of plants for the chemical conversion of raw materials into marketable products which, as commodities, met the specifications for all means of transport powered by internal combustion engines. Purchasing departments have been set up to procure the required quantities of raw materials – plus the quantities needed to compensate for processing losses of between 10% and 20%, depending on the process. The make-or-buy decision was easy because suitable vegetable oils and animal residues were available, and buying the raw materials was cheaper and easier than setting up a vegetable oil production through agriculture. The additional demand for vegetable oils exacerbated the existing volatility of the vegetable oil market, whose price movements had previously been largely in line with available crop yields.

The separation of raw material production and raw material processing on the capital side followed the plausible approach that the technical requirements of agricultural production are clearly different from the technical requirements of a chemical processing plant. The only interface would have been the advantage of a single-stage process chain, which would have allowed optimisation between agricultural production and the required processing volumes at predictable prices. As long as only a few buyers faced a wide range of supply, concentrating capital investment on production facilities was the right business decision. The diversified demand for vegetable oil products by individual SAF producers has not yet led to a measurable increase in the area under vegetable oil cultivation, especially as annual oil crops have to be rotated to avoid soil degradation in individual fields. Perennial crops such as palm fruit or Jatropha curcas require several years of growth to produce usable yields. The low and sometimes irregular demand for energy purposes has therefore not been an incentive for the capital market to expand the investment portfolio to include perennial oil crops, especially as their geographical cultivation is concentrated in subtropical regions where different country risks have to be taken into account.

A constant supply leads to price competition for the available quantity when demand increases and thus to price increases as long as there are no substitutes to offset the excess demand. In the example of SAF's production from vegetable oils, the buyers' search for substitutes focused on the market for used cooking oil (UCO), which is regularly generated as waste in the food industry and catering sector. Although UCO is only available in small quantities and has to be collected locally, this waste material is attractive for use as a raw material in the energy sector because it is not problematic from an environmental point of view and is also produced in almost constant quantities. However, the increasing demand for UCO with a constant supply also leads to an excess demand, which is reflected in rising prices. According to media reports, the price of UCO has repeatedly exceeded the price of fresh cooking oil. Some sellers were able to sell artificially inflated quantities of UCO by adding fresh cooking oil, which was of course illegal (Argus Media, 2020; Euractiv, 2021).

The coming global demand for bio-commodities for the production of diesel and jet fuel can only be met by a massive expansion of arable land. In countries with high cereal exports, especially the US, Canada and Ukraine, a large amount of arable land can be converted to growing cereals specifically for energy production. From a macroeconomic point of view, this conversion will take place autonomously by producers when the return from energy commodities is higher than the return from food grains. Such a shift towards food production will affect the food supply in many countries, especially as many developing countries are dependent on food imports. In a broader context, there is still a lack of strategy on how to implement the substitution of fossil fuels with synthetic fuels, given the availability of large quantities of biomass-based feedstocks. The forthcoming mandatory blending quotas for SAF in the EU and the announcement by IATA that airlines will offset 65% of aviation emissions with SAF by 2050 are not yet backed up by the volumes of feedstock required. It is expected that the demand for SAF will grow faster than the construction of new production facilities. As a result, the supply of feedstocks for the new synthetic fuel production facilities will become a consequential problem, as the expansion of suitable biogenic feedstocks will also lag behind demand. If there is an undersupply in the market, not only will feedstock prices rise – exacerbating food availability – but investment in new production facilities will also be put on hold if the feedstock supply for plant operation is not guaranteed. However, a backward-looking process chain cannot avoid upstream bottlenecks. When the crude oil market developed, the feedstock was available and suitable outlets were sought. Looking backwards from the demand side to the source of the raw material carries the risk of competitive reactions. If competitors with RFNBO feedstocks are able to enter the market faster and earlier than the development of the biogenic feedstock chain can be implemented, the bio-based SAF production pathways can only participate in the market growth with a time lag and only at the market prices that have been established in the meantime. A raw material supply shortage therefore not only affects the business model of an individual investment in an SAF production plant, but can also become a problem for an entire industrial sector if the raw material supply is dependent on imports and corresponding cultivation areas have to be developed first.

A competitive situation also arises within the SAF market with regard to the use of feedstocks: (1) The intrinsic competitive situation of SAF in relation to the larger diesel market described above leads to the use of feedstocks in diesel production if this becomes more profitable than SAF production. (2) Furthermore, it must be determined in advance for which geographical market the fuels are to be produced. Different regulatory frameworks in the US and Europe have an impact on the sourcing of feedstocks. (3) Given an existing range of different feedstocks, it must be determined which of the feedstocks are suitable for the production process, which are available throughout the calendar year and what additional pre-treatment may be required. (4) From the point of view of security of supply, the transport distance from the so-called "first gathering point" of the raw material to the production plant, as well as an unavoidable change in the means of transport or any other transport risk, may cause an unexpected interruption in the transport chain and lead to a shortage of supply. (5) An indirect shortage of raw materials occurs when the raw materials are available but are diverted to another market due to higher selling prices. In addition to the transport sector, the chemical industry is also a large demand group. In terms of the end product price, the chemical industry will be able to accept higher raw material prices for its production than the transport sector. For SAF feedstocks, this creates an arbitrage situation in which SAF producers must at least meet the purchase prices for feedstock supplies set by the chemical industry in order to gain or maintain access to biomass-based feedstocks. The risk of increasing production costs associated with individual SAF production processes cannot be mitigated to the extent necessary to achieve the planned return on capital employed. If affected by arbitrage pricing, investments in new plants are at risk of being unprofitable and existing plants may be prematurely shut down. (6) Finally, the frequency of labour disputes along the upstream process chain for raw materials must also be included in the consideration of possible risks.

It can be concluded that the risk of supply disruption increases with the complexity of the process chain.

The circumstances described here for raw material supply problems are only intended to illustrate the complexity of SAF's process chain. All the sources of disruption mentioned are basically manageable, but they require a comprehensive and careful analysis of all factors. The basic problem is the uncertainty factor in assessing raw material properties, availability, regularity of supply and the likelihood of disruptions that originate outside the raw material supply chain but affect the stability of supply. The assessment of uncertainty in investment decisions is a recurring problem in economic analysis and is not limited to the variables of SAF production.

The procedures presented below are intended only to provide a general overview of the options for dealing with uncertainty:

1. Correction method:
   A simple method of dealing with uncertainty is to correct the economic indicators by a fixed or percentage addition or deduction based on the assessment of each factor. However, this method is subject to subjective judgements and therefore has limitations as a tool.
2. Sensitivity analysis:
   Sensitivity analysis is used to change variables that are relevant to the outcome in such a way that their effect on significant factors can be

represented and measured. It allows the range of possible outcomes to be identified using different scenarios. At the same time, sensitivity analysis provides the maximum and minimum expected results ("best case/worst case"), provided that the range of changes in the earnings-relevant variables is correctly represented.

3. Risk analysis:

Risk analysis extends the methodology of sensitivity analysis to include the evaluation of individual conditions and scenarios with their respective probability of occurrence. Simulation calculations are used to link the level of results to the associated probabilities of occurrence.

By assessing the profitability of an investment, the uncertainty of results can be reduced.

4. Decision tree analysis:

In anticipation of future changes in environmental legislation and their impact on the profitability of long-term investment decisions, exogenous factors should also be included in the assessment of profitability. Exogenous changes that will affect production conditions in the future are given probabilities of occurrence and presented in a decision tree by period. The decision tree visualises opportunities and risks with effects on the break-even point, the payback time and the expected return.

5. Value in use analysis:

The benefit analysis expands the scope of the investment to include factors that may indirectly affect the return on the investment. In the present case of feedstock supply for SAF production, these factors include the change in social values towards environmentally friendly and resource-saving production, the use of renewable energy as an additional benefit for climate protection, and thus the question of future consumer orientation and the willingness of consumers to pay a premium for environmentally friendly products.

Conclusion: The necessary expansion of SAF production is not only based on the accelerated construction of production facilities, but must also take into account the availability of raw materials, their price development and their creditability in emissions trading systems.

4 Investment Considerations

The net-zero 2050 sector target is for 63% of aviation emissions to be offset by SAF. The remaining 37% will be covered by offsetting (purchase of allowances or abatement investments in other sectors) and technological progress in propulsion technology and air traffic management. With regard to expected technological progress, the aircraft and engine manufacturers do not provide any clear justification as to which innovations could lead to a corresponding reduction in emissions and how widespread these innovations will be in the global aircraft fleet in 2050. While the US has already modernised its airspace with the FAA's NextGen (Next Generation Air Transportation System) programme using Trajectory Based Operations (TBO), in Europe the SESAR (Single European Sky ATM Research) project is still awaiting implementation in air traffic control in European airspace. The focus of emission reduction is therefore on SAF rather than on operational performance improvements. It should be noted that the percentage of SAF use varies from 63% to 70% in publications and press releases, while the base fuel volume on which such proposed or projected figures are based remains unnamed.

From an investor's point of view, the question of the availability of alternative propulsion systems in aviation for the period 2030–2050 needs to be answered, especially as considerable research and venture capital is being spent on environmentally friendly propulsion technologies. Electric and hydrogen propulsion are mentioned as alternative energy sources. For both alternatives, research is still at a stage where only experimental aircraft with small payloads could fly over short distances. Both technologies face challenging tasks, as research in both cases must overcome technological barriers that have not existed to the same extent in other areas of technology. Aircraft require fuels with high energy density and low specific weight. The development of electric propulsion for aircraft is feasible, but the electricity has to be stored in batteries, which are heavy and have a low storage density. Technological progress has increased the performance of batteries by 100% over the last 20 years. This increase is not sufficient for efficient aircraft propulsion. The energy density of batteries needs to be increased by a factor of 40 while keeping weight to a minimum. Aircraft in regional traffic could be converted to hybrid propulsion for short distances by 2050, but the resulting fuel savings are rather small. Hydrogen propulsion is conceivable for aircraft. However, due to its lower density than JET A-1, hydrogen requires a large pressurised cylinder tank, which must also be pressure and temperature resistant in order to transport liquid hydrogen at −160°C. For short distances, the rear section of the aircraft cabin could be converted into a tank, but this would make the aircraft extremely uneconomical due to the loss of earning capacity (revenue losses).

DOI: 10.1201/9781003440109-4

For both alternative energy sources, the airport supply infrastructure would have to be completely rebuilt. Due to their performance parameters, electric and hydrogen-powered aircraft are also likely to require longer runways in order to reach the required take-off speed. Lengthening runways by up to 25% is not feasible at many airports due to lack of space, so hybrid aircraft will probably only be able to take off at airports with existing long runways. This is not the case at many regional airports. Against this background, it is very unlikely that alternative forms of energy will limit the demand for SAF by 2050. Investments in SAF production facilities and in the supply of raw materials for SAF can be considered safe, at least in terms of volume demand.

The conversion of JET A-1 from fossil crude oil to climate-friendly synthetic fuels will take at least until 2050 and will require immense investments in the creation of the necessary production facilities as well as in the further breeding and cultivation of energy crops or in the provision of "green" electricity as an energy source for the production of hydrogen or as a direct energy source for electric drives. Based on a demand volume of 300 million tons of JET A-1 in 2019 and an expected increase in traffic to around 500 million tons in 2050, civil aviation and the ICAO's "net zero" approach assume that 63%–70% of the JET A-1 demand will be covered by SAF by 2050. In this context, SAFs are still calculated as "100% emission-free", although in reality their effective emission reduction is only between 70% and 95%. Under these assumptions, a net production capacity of $500 \times 0.7 = 350$ million tons will be required for SAF production by 2050. Cost-optimal co-production also requires the production of short-chain hydrocarbons. Longer-chain hydrocarbons such as diesel and heavy fuel oil can be incorporated into the kerosene fraction through post-treatment, but the synthetic kerosene yield from any of the approved production processes will not exceed 90% on average due to unavoidable production losses.

To keep the example calculation of investment requirements for eSAF production transparent, the production loss is not added to the plant capacity. However, the raw material input is increased to compensate for the loss. This is relevant for the determination of the production costs, but not for the determination of the plant investment volume.

Plant sizes and production capacities vary from region to region. For simplicity, an average plant size of 800,000 tons of SAF jet fuel production and 100,000 tons of by-products per year is assumed. The investment sum is a flat rate of USD 1 million per 1,000 tons of production capacity, i.e. USD 900 million per unit.

For the plant to operate at full capacity, between 1 million tons (vegetable oils, HEFA pathway) and 9 million tons (woody residues, municipal solid waste, FT-SPK pathway) of raw materials will need to be sourced and transported to the production site. The investment in the feedstock supply chain for HEFA-SAF in this example calculation is USD 0.2 million per 1,000 tons of feedstock produced. These include, for example, farm vehicles, warehouses, silo facilities, oil mills (extraction and mechanical pressing), harvesters, tank facilities and transport vehicles. In the case of RFNBOs, the estimated investment in power supply (wind and solar) and electrolysis is USD 6 million per 1,000 tons of production capacity, which is significantly higher than the investment in competing SAF production facilities.

A payback period of 20 years per plant is assumed for the SAF production facilities and for hydrogen production. For the biomass feedstock supply chain, an average depreciation period of 10 years can be assumed.

In order to theoretically replace 70% of the JET A-1 net volume of 300 million tons in 2019 (210 million tons) with SAF, it is necessary to convert or build new production facilities.

It is further assumed that the availability of certified biomass as an SAF feedstock is limited globally and that only 50% of the SAF demand can be met by biomass-based SAF. The remaining 50% of the demand will have to be met by the so-called eSAF (power-to-liquid SAF).

For the construction of a HEFA refinery with an output of 800,000 tons of SAF and 100,000 tons of co-products, the total investment required for the plant and related investments in the feedstock supply chain is estimated at USD 1.1 billion. For AtJ and FT-SPK (waste-to-liquid SAF), the investment in this model calculation is of a similar magnitude. For eSAF, a much higher investment is required. Hydrogen production from renewable electricity by electrolysis requires investment in electricity generation based on wind and solar farms as well as the design and construction of appropriately sized electrolysers. A conservative estimate of the investment required for a (virtual) 800,000 ton eSAF plant with an additional 100,000 tons of by-products is USD 3.0 billion, and a further USD 3.0 billion for gas-to-liquid synthesis gas production followed by Fischer–Tropsch synthesis.

In this initial model, which reflects the demand situation in 2019 and a theoretical SAF production rate of 70% (210 million tons), 263 units are required, of which 131 are HEFA refineries and 132 are eSAF refineries. The capital requirement for this baseline would have been USD 909.9 billion at 2023 prices.

In the target year 2050, 70% of the net 500 million tons of JET A-1 (350 million tons) would have to be replaced by SAF and eSAF. Out of a total of 438 plants, 219 are covered by biomass-based plants and 219 by eSAF from hydrogen and a natural carbon source. The capital requirement for the 438 production facilities, including the provision of "green" electricity, is USD 1,511 billion at 2023 prices.

For the sake of completeness, the scenarios also consider 100% emissions offset by SAF and eSAF in 2050. This extreme case is unlikely, but with insufficient CO_2 offsets and little technological progress in engine and aircraft design, only a change in fuel type remains as a means of actually achieving "zero" rather than "net zero" emissions. Again assuming 50% biomass-based SAF and 50% eSAF for this scenario, the capital requirement for 625 production plants is USD 2,221.2 billion, with an SAF/eSAF output of 800,000 tons per plant and 100,000 tons of by-products per plant. Based on a demand volume of 500 million tons, the debt service (20-year depreciation and average return on capital at a 9% interest rate) is approximately USD 337.6/cbm or USD 1.28/USG at 2023 prices.

Whether this theoretical model will be applied in reality can only be determined to a limited extent by means of a probability calculation. However, if the emissions balance of aviation does not reach the level required to mitigate global warming, an internationally coordinated limitation of aviation emissions through the apportionment of CO_2 and non-CO_2 emissions by limiting the permissible TKO (ton-kilometres offered) could be a realistic alternative.

4.1 INVESTMENT IN RAW MATERIALS

In order to keep the example calculation of investment requirements for SAF production transparent, the production loss is not added to the plant capacity. However, the feedstock input is increased to compensate for the loss. This is relevant for the determination of the production costs, but not for the determination of the plant investment volume.

The following conversion processes are suitable for the production of SAF:

1. thermo-physical conversion processes based on cellulosic raw materials,
2. physical–chemical conversion processes based mainly on vegetable oils, animal fats and algal oils, and
3. biochemical conversion processes using raw materials for the production of intermediates by alcoholic fermentation of straw and sugarcane to ethanol/isopropanol or anaerobic fermentation of biomass and manure to biomethane.

4.1.1 FUNDAMENTAL CHALLENGES OF INVESTING IN NATURAL RESOURCES FOR SAFs

It is much more difficult to invest in the cultivation of raw materials than in the establishment of a waste collection and recycling chain. There are several reasons for this: (1) Agricultural biomass is perceived as risky because harvest volumes vary seasonally, and longer-term contracts with buyers must therefore take into account different unit costs depending on harvest volumes. Although the price adjustment based on the harvest is necessary to ensure the existence of the farm, it makes it difficult to conclude multi-year supply contracts, as the quantity uncertainty cannot be compensated by price advantages. (2) In addition to weather-related fluctuations in crop yields, pests and fires can also reduce crop yields. (3) For certified crops, the use of artificial fertilisers is limited or even prohibited. This also applies to any form of artificial irrigation. (4) Harvested crops can only be stored for a limited period of time. Especially in the case of oleaginous fruits, the oil content decreases continuously during storage, so that rapid processing is necessary for yield reasons. (5) Depending on the biomass, the harvest period is limited to a few weeks per year. Extensive machinery must be available for the harvesting period, which is only used for a limited time. The same applies to warehouses, silos and tanks where the harvested material must be temporarily stored until it is sold. For oilseed crops, extraction plants or oil mills for mechanical pressing are also required. (6) The predominant farm sizes in Europe are labour intensive and can only maintain a limited range of machinery for economic reasons. European farms have to be supported by agricultural subsidies from the Member States of the European Union. Given these financial dependencies, there remains uncertainty about the eligibility of energy crop cultivation for subsidies. (7) Annual crops, i.e. energy crops, usually require 3–4 years of crop rotation to maintain soil quality. In this respect, not only energy crops can be grown, but also co-rotating crops whose markets are different from those of energy crops. Perennial energy crops – particularly oilseed crops – require an initial growing season of several years before they start to yield. During this period, the seedlings require care and protection from pests. The time between providing the cash to establish a new acreage and matching a break-even point can be as long as 7–9 years. This is also critical from an investment point of view, as a rapid return on capital promises greater investment security. (8) From an investor's point of view, the American farm model of large acreages with appropriate personnel, technical equipment and a high use of machinery in field work is therefore an important indicator of success. (9) In order to establish new farms with the characteristics described above, land must be developed in temperate and subtropical climates where soil quality and rainfall promise successful cultivation. Although it is possible to use degraded land, the quality of the land has a direct impact on crop yield and thus on the profitability of the investment. (10) When land is developed in Southern Africa, Central America or South-West Asia, there is a lack of the necessary infrastructure, such as roads that can be used all year round, bridges, energy supply and a skilled workforce that is familiar with agricultural practices and has the appropriate skills. (11) In addition, the preparation of land for agriculture ("land clearance") is costly and time-consuming. (12) In politically unstable countries, the risk of civil unrest, rebellion and civil war is an additional investment risk. (13) Stable import conditions in industrialised countries are essential for the export of raw materials. For example, the constant amendment of the European Union's Renewable Energy Directive is an obstacle to investment, as the conditions for agricultural cultivation change over time, making it difficult or even impossible to export biomass.

(14) Last but not least, the recurring discussion on land use change (LUC), indirect land use change (iLUC) as a result of predatory competition with the reclamation of previously unused land, as well as the discussion on food security in developing countries have so far prevented the expansion of energy crop cultivation outside existing cultivation areas. A life cycle analysis (LCA) of the entire supply chain, from the cultivation of the raw material to the sale of the refined SAF product, is required as a prerequisite for emission reduction certificates that can be used as a substitute for the purchase of CO_2 certificates in the EU-ETS or as a qualification in accordance with EPA requirements for RINs (Renewable Identification Numbers) as tax credits in the US. Accredited certification schemes are ISCC (International sustainability and carbon credits) and RSB (Roundtable for sustainable bio-materials).

4.1.2 FEEDSTOCKS FOR THERMO-CHEMICAL PROCESSING PATHWAY FT-SPK

In the gasification pathway, biomass is converted to synthesis gas in a reactor in the absence of oxygen. This technology received its first SAF approval in 2009 as FT-SPK. At that time, the anticipated feedstock was residual wood and wood chips from short-rotation plantations such as poplar and willow. So far, woody feedstocks have not proven to be a viable biomass energy source: (1) The low energy content of cellulosic feedstocks results in a disproportionate raw material input of 8:1, i.e. 1 ton of fuel requires 8 tons of biomass to be gasified. (2) Although the woody raw materials are cheap, their collection and transport is very costly. Residual wood from forests must be collected and transported to a first gathering point, often over unpaved roads. From there, the wood is loaded onto trucks. (3) If too much residual wood is removed from an intact forest stand, the forest lacks fertilisation from the decomposition of the residual wood. This must then be replaced by artificial fertiliser, which is an additional cost. (4) There is a risk of soil degradation in short rotation plantations. Even if the woody plants can be harvested mechanically, this method of producing raw materials requires a great deal of land. (5) There is an interdependence between the location of the production plant and the supply of raw material from neighbouring forests, as the large quantities of harvested material or residual wood cannot be transported over long distances, as this would make the transport costs disproportionately high, and at the same time, the transport emissions would worsen the LCA. In addition, short-rotation plantations require a growth period of 5–7 years before they are cut or cleared. (6) The use of energy grasses such as switchgrass is also subject to similar restrictions, so that project-related cultivation of these raw materials in the context of an SAF gasification project has not yet taken place. (7) In particular, the increasing risk of forest fires and the infestation of trees by insect pests represent further risks. (8) As a raw material with low energy content, the return on investment for such projects is also low.

Because of the problems with biomass cultivation described above, research in the early 2010s focused on the use of woody waste materials. Until then, there was a little-known relationship between the composition of the biomass and the complexity of the gasification process. The more heterogeneous the biomass composition, the more complex the gasification process. This meant that certain types of biomass had to be thermally pre-treated – through what is known as torrefaction or low-temperature pyrolysis – in order to achieve an acceptable result in the subsequent gasification process. Torrefaction increases the size of the plant, requires additional investment and adds to the cost of producing syngas. As a result, this process route has not been able to establish itself either technologically or commercially. The technological demands on the gasification reactor are high, and the conventional temperature range limits the production of commercially viable syngas.

The development of high-temperature gasification has at least made it possible to increase the performance of the gasification reactor. This has made the technology relevant for other waste materials. In many countries, municipal solid waste (MSW) is disposed of in landfills or incinerated in waste-to-energy plants. The heat from incineration is used to generate electricity via a steam turbine, but the flue gases need to be cleaned to prevent additional air pollution. At the same time, the incineration process produces additional CO_2, which is released into the atmosphere. The energetic and at the same time environmentally friendly use of MSW is currently the focus of the FT-SPK route for SAF. MSW is an energy-rich waste material if it has not been previously sorted for recyclable materials, which makes the residual energy content of MSW critical for gasification. In metropolitan areas, large quantities of household waste are generated regularly, are available regardless of the season and can be reliably delivered to SAF's production plants if they are collected on a decentralised basis. However, the gasification of household waste requires the separation of metals and, if necessary, the shredding of the waste into small pieces.

4.1.3 Feedstocks for Physical–Chemical Processing Pathway HEFA

HEFA (Hydrotreated Esters and Fatty Acids) is currently the SAF process with the highest technology maturity (TRL 9) and the largest SAF output on a global scale. In the start-up phase following ASTM approval in 2011, palm oil from certified cultivation (RSPO Roundtable for Sustainable Palm Oil) was used as raw material alongside other vegetable oils. After environmental organisations opposed the energetic use of palm oil in the context of the food versus fuel discussion, the energetic use of this type of oil was discontinued. Instead, from 2013 onwards, intensive consideration began to be given to the oil plant Jatropha curcas, which was seen as a beacon of hope for energy crop cultivation and was to be grown as a perennial plant in a plantation economy. Jatropha curcas was considered a frugal plant that could survive dry periods with its taproot and could be planted out on semi-arid soils after previously growing the seedlings. With the help of venture capital, numerous plantation projects were started in Mexico, Africa, Madagascar, India, Indonesia and the Philippines. Wild plants were planted as there should be no delay in establishing a supply chain based on Jatropha oil in the absence of a lead time for breeding suitable seeds. The observed "Jatropha hype" was done with the aim of investing available investment funds in farm development as quickly as possible. In doing so, all requirements of sustainable and science-based field management were disregarded in order to be able to show investors the most rapid "quick wins" possible in order to then be able to call up further tranches of investment funds. The mistakes made can be easily understood – without claiming to be exhaustive – even by non-experts: (1) cultivation on degraded soils whose fertility was not sufficient to stimulate plant growth and thus the yield of oleaginous nuts; (2) no analysis of weather data and therefore cultivation on land with too little rainfall and a too lengthy dry season. If the dry season exceeds 3–4 weeks, the plant sheds all its leaves and does not produce nuts in the crop year; (3) establishment of farms by non-expert managers without taking into account the position of the sun and prevailing wind direction; (4) use of seeds from wild plants with low yield; (5) lack of establishment of bees and other insects for pollination; (6) lack of knowledge of how to prune annual shoots in the growing season to achieve maximum twining, which is essential for flowering and nut germination; (7) unsuitable agricultural equipment for weeding, as wild weeds extract nutrients from the soil, which are then no longer available to the cultivated plant; (8) theft of fertiliser and its sale on the black market instead of the legal and proper fertilisation of the plant as required by certification; (9) employment of unskilled labour with no experience in handling the Jatropha plant; (10) lack of investment in storage facilities and oil mills; and (11) inadequate planning for the transport of the plant oil by truck or train and for the use of the husks as fertiliser and the press cake as burning material for local use instead of the use of charcoal, which contributes to deforestation.

This extensive list serves to explain the reasons for the failure of Jatropha cultivation in developing countries. For these countries, successful Jatropha cultivation could have become a sustainable and environmentally friendly economic factor, creating jobs in rural areas and preventing rural depopulation. For the aviation sector, the vegetable oil, which is not suitable for human consumption, would have become an important source of energy for the decarbonisation of air transport, while the plant proteins, once detoxified, would have been a cheap alternative to soya meal as animal feed in these countries (von Maltitz et. al., 2018).

After the failure of Jatropha cultivation, research and development focused on the extraction of algae oils. Algae oil is a valuable raw material in the cosmetics industry. The US and Australia have led the way in the development of large-scale algae cultivation due to their climatic advantages of high solar radiation and warm ambient temperatures. In this area of research, too, various projects have been funded with venture capital, but without producing results that could be used in the transport sector: (1) algae cultivation in open pond systems failed due to the ingress of pollutants and bacteria from wind and bird droppings. (2) Algae cultivation in closed reactor systems (glass bodies) required artificial lighting of the containers. (3) As algae increasingly colonise the illuminated glass walls, this colonisation leads to a reduction in light input in the centre of the reactor, which is accompanied by a reduction in algae growth. (4) The bright lighting in the cultivation reactors shortens the lifespan of the algae strains to a few weeks, requiring additional expenditure for permanent cultivation. (5) The CO_2 source required for algae growth must ideally be located in close proximity to the algae cultivation to avoid long transport routes. (6) Separating the algae oil from the water phase is costly and technologically complex; and (7) with today's processing technologies, large-scale production of a continuously available algae oil for air transport has not been achieved.

Annual vegetable oils such as rapeseed or Camelina sativa are currently used for HEFA production. Non-food palm kernel oil is also increasingly being used. In the US and northern Europe, animal fats are also obtained from slaughterhouse waste. The range of oils and fats used as energy feedstocks has increased.

In the context of the use of household waste as an energy source in OECD countries, a new use for used cooking oil (UCO) from food processing and restaurant kitchens has also developed. These are mainly frying oils and fats, which initially had to be disposed of as waste. After appropriate filtration, these oils could be blended with biodiesel made from rapeseed oil (FAME – Fatty Acid Methyl Esters). It has now been shown that UCO can also be used as a feedstock in the HEFA jet fuel process. As with all household waste, UCO is produced regularly in roughly constant quantities, can be collected locally in metropolitan areas and transported to refineries in tankers or rail tank cars.

4.1.4 Feedstocks for Biological–Chemical Processing Pathways ATJ and Biomethane-PtL

As an alternative to the FT-SPK route, the use of herbaceous biomass for ethanol production has emerged as a suitable feedstock along with industrial wheat, maize and sugar beet for ethanol production in temperate climates. In tropical and subtropical climates, ethanol is produced from sugarcane and the processing of bagasse residues. While ethanol as a blending component for gasoline is already a marketable product (a so-called "intermediate"), ethanol and the related isobutanol can be further processed into SAF by adding a second conversion step. The alcohol-to-jet feedstock pathway is thus also based on cultivated biomass but follows a biochemical conversion pathway instead of gasification.

The biomethane-to-GtL route and the use of hydrogen and carbon from a natural source followed by gasification and Fischer–Tropsch synthesis represent a combination of the use of biogenic feedstocks and an RFNBO route from hydrogen and biogenic methane gas. Technologically, they are FT-SPKs, but their anaerobic fermentation provides a bridge to the biochemical conversion processes. Biomethane is obtained from the fermentation of biogenic waste, chicken manure and sewage sludge. The raw biomethane gas can be burned as an energy source in modified internal combustion engines, driving a generator to produce electricity. The process heat is used to heat greenhouses or public facilities (e.g. swimming pools). In biomethane production, only the energy is extracted from the substrate; the digestate can then be used as fertiliser in agriculture.

In the gas-to-liquid route, synthesis gas is produced in reactors. Depending on the design, different feedstocks are used. Non-biogenic feedstocks include coal and natural gas, as well as hydrogen from electrolysis using green electricity from wind, solar or hydro power. For the production of SAF, coal and natural gas must be excluded as they are fossil fuels that should not be used in the future.

While biogas plants tend to be "stand-alone" and are financed by farms, hydrogen production from electrolysis and the associated investment in solar and wind farms will become more important in the future. Like the biogenic SAF pathways, the PtL route separates the production of raw materials (green electricity and subsequently hydrogen) from the use of hydrogen to produce synthetic fuels. Due to the direct coupling of electricity and hydrogen on the one hand, and the direct use of hydrogen as a feedstock for the production of power-to-liquid jet fuel on the other, cross-sectoral capital provision is advisable in this case.

4.1.5 Feedstock Investment Outlook

The use of biomass as a feedstock for SAF between 2009 and 2023 shows different stages of feedstock development from cultivated biomass to the use of waste materials of biogenic and non-biogenic origin. For economic reasons and to minimise harvest risks, cultivated biomass is currently grown on existing oilseed and cereal cropland, with annual crops dominating. The cultivation of perennial oil crops is limited to South-East Asia. Here, the by-products of palm oil production are used for residual purposes without affecting the use of palm oil itself as a raw material for food and animal feed.

Furthermore, the basic segmentation of SAF production into agricultural feedstock production, including feedstock processing (such as the extraction of vegetable oils by extraction or mechanical pressing, and the production of "intermediates" such as ethanol), is more likely to be attributed to the agricultural sector, even if chemical processes are used in the process. On the other hand, the industrial production of synthetic diesel and SAF belongs to the chemical sector and its large-scale industrial production plants. Both sectors are subject to different framework conditions, and therefore, each sector has its own specific investment requirements. In this respect, it should be noted that since the approval of the FT-SPK process route in 2009, no cross-sectoral integrated SAF production from feedstock production to final product has been established, and therefore, there is no integrated investment decision that fully maps the production chain. To date, the supply of raw materials has not proved to be a bottleneck in SAF production. This is either due to the lack of market maturity of an individual SAF production process, which is the main reason for the lack of scale, or to the satisfactory availability of raw materials in a polypolistic, price-sensitive supply market, which provides SAF producers with sufficient options to choose from and at the same time spreads the risk in terms of security of supply.

With an increasing number of SAF production facilities and an almost unchanged supply of suitable feedstocks, the point will be reached in the future where a real shortage of raw material availability will set in. Even before this – virtual – point is reached, the prices of these feedstocks will rise since, in any market model, excess demand leads to a price increase to absorb the consumer rent under the premise of profit-maximising production.

As long as there is a significant technology-driven difference in production costs between biomass-based SAF production processes and RFNBO-SAF production plants, the currently known pathways and their feedstocks will remain attractive to investors, provided that the technology has already reached a high degree of maturity and the supply of raw materials must be secured in the medium to long term.

If raw material shortages are expected over a longer period of time, it should be considered on a case-by-case basis whether it is economically advantageous to integrate the raw material supply into the process chain of the production plant. When a polypolistic market turns into an oligopoly, securing the availability of production factors at favourable costs is an alternative worth considering – at least for the proportion of raw materials required to minimise the risk of losses, or even the risk of production cuts due to unavailability of raw materials or raw material prices that are not affordable to continue full production.

For future investments in the feedstock supply chain, it makes sense to analyse the success factors and preconditions of US agriculture, paying particular attention to the expansion of agricultural land. The mistakes made in jatropha cultivation and the problems encountered in algae cultivation show that such projects require much more intensive preparation and a comprehensive risk/reward analysis. In addition, the quality and experience of the management of such projects is a critical success factor.

Conclusion: From a risk perspective, investments in the cultivation of energy feedstocks should not focus exclusively on the "SAF" pathway. Diversification of cultivation for different demand groups and customers spreads the risk and prevents a situation that could threaten the continuation of the business if, for whatever reason, SAF producers no longer demand the agricultural product supplied by the recipient of the investment.

Depending on the maturity of energy crop cultivation, investments in further crop development could be helpful for the supply of feedstock for SAF if higher yields can be achieved through further developed seeds or if the energy crop becomes more suitable for a wider range of habitats, which would also improve the supply of feedstock for SAF production.

4.2   INVESTMENTS INTO SAF PRODUCTION FACILITIES

SAF – Production facilities are at the centre of investment considerations. Based on a demand forecast of SAF volumes over the amortisation period of an industrial plant, it is possible to plan, finance, build and operate a balance sheet-compliant plant. In addition, such a plant has the advantage that the feedstocks can be sourced from a variety of sources, thus ensuring security of supply and a reasonable price for the raw materials.

Since the beginning of oil production, fuel supply has been in the hands of international oil companies and, in some countries, national oil companies. Decades of experience in handling hydrocarbons and the well-to-wheel logistics chain have given the oil companies not only extensive expertise but also knowledge of the markets and their characteristics.

The approval of synthetic fuels for ground transport and aviation opens up opportunities for new companies to enter the market for the first time. This is particularly true for SAF, even though it is currently only approved as a sustainable blending component (SBC). Aviation fuels are commodities, i.e. products with identical specifications that can be used regardless of the manufacturer if the product is within the specification limits in all given parameters. There are various refining technologies and processing methods to produce fuel from crude oil. Crude oil is a raw material with a wide range of properties that must be used in the refinery in such a way that the end result is a product that conforms to the specification.

As far as SAF is concerned, the Finnish oil company Neste OY was the first oil company to build its own refineries for the production of synthetic fuels in addition to its crude oil refineries, and since 2010, it has also been producing "NexBtL" (now "HEFA") in these plants, at least in batches. Meanwhile, Neste has technically upgraded some of its plants and is able to produce SAF at any time in the quantities required by the market. Years later, ENI (Italy) and TOTAL (France) followed this trend in Europe by converting old mineral oil refineries to HVO/HEFA refineries. In the US, the conversion of the former Paramount petroleum refinery (California) into a HEFA-producing refinery, currently operated and owned by World Energy, was the first step in 2018 to produce SAF in combination with other fuels.

However, the converted refineries in Europe primarily produce synthetic diesel, which is in demand as a premium product in Europe and therefore offers attractive profit margins. With the introduction of aviation blending mandates in Norway and Sweden, the previously negative attitude of European legacy carriers towards SAF has changed. For European airlines, offsetting their greenhouse gas emissions is a more economically viable option than purchasing SAF volumes equivalent to the required CO_2 emission reduction certificates. Against this background, SAF sales in Europe remained below expectations. Only small volumes were used in connection with EU-funded research projects, while the Dutch airline KLM was able to convince corporate customers to offset emissions from business travel through an incentive programme. The SAF quantities that could be financed were purchased in the US and used on flights from US territory to Amsterdam.

While the national governments of the EU member states were waiting for a pan-European approach to reducing emissions from European aviation, the US government pursued a different policy from the beginning. Based on the California state government's Low Carbon Fuel Standard (LCFS), which has limited emissions from road transport since the 1990s, the federal government has relied on subsidy programmes to encourage the purchase of biofuels. This approach has been very successful to date, and it will receive further impetus to decarbonise road and air transport in 2022 with the Inflation Reduction Act. It is therefore plausible that, with the exception of the HEFA process for jet fuel, all new renewable fuel production technologies have been developed and submitted to ASTM for approval in the US, Japan and South Africa in recent years. As production processes continue to improve, the scale-up of new technologies to large industrial plants is accelerating.

There is no doubt that the international aerospace industry will significantly increase its demand for SAF in the very near future due to the mandates and incentives mentioned above. It should be borne in mind that compliance with the blending mandates is the responsibility of the jet fuel producers, who are obliged to pay penalties to their tax authorities in the event of non-compliance. Conversely, airlines are obliged to offset their emissions in the European Emissions Trading Scheme and require emission reduction certificates from their suppliers' mandates. In the US, incentives are provided to both producers (US Blenders Tax Credit) and purchasers of the SAF (RINs). As far as SAF is concerned, the Finnish oil company Neste OY was the first oil company to build its own refineries for the production of synthetic fuels in addition to its crude oil refineries, and since 2010, it has also been producing "NexBtL" (now "HEFA") in these plants, at least in batches. Meanwhile, Neste has technically upgraded some of its plants and is able to produce SAF at any time in the quantities required by the market. Years later, ENI (Italy) and TOTAL (France) followed this trend in Europe by converting old mineral oil refineries to HVO/HEFA refineries. In the US, the conversion of the former Paramount petroleum refinery (California) into a HEFA-producing refinery, currently operated and owned by World Energy, was the first step in 2018 to produce SAF in combination with other fuels.

However, the converted refineries in Europe primarily produce synthetic diesel, which is in demand as a premium product in Europe and therefore offers attractive profit margins. With the introduction of aviation blending mandates in Norway and Sweden, the previously negative attitude of European legacy carriers towards SAF has changed. For European airlines, offsetting their greenhouse gas emissions is a more economically viable option than purchasing SAF volumes equivalent to the required CO_2 emission reduction certificates. Against this background, SAF sales in Europe remained below expectations. Only small volumes were used in connection with EU-funded research projects, while the Dutch airline KLM was able to convince corporate customers to offset emissions from business travel through an incentive programme. The SAF quantities that could be financed were purchased in the US and used on flights from US territory to Amsterdam.

While the national governments of the EU member states were waiting for a pan-European approach to reducing emissions from European aviation, the US government pursued a different policy from the beginning. Based on the California state government's LCFS, which has limited emissions from road transport since the 1990s, the federal government has relied on subsidy programmes to encourage the purchase of biofuels. This approach has been very successful to date, and it will receive further impetus to decarbonise road and air transport in 2022 with the Inflation Reduction Act. It is therefore plausible that, with the exception of the HEFA and HC-HEFA processes for jet fuel, many of new renewable fuel production technologies have been developed and submitted to ASTM for approval in the US in recent years. As production processes continue to improve, the scale-up of new technologies to large industrial plants is accelerating.

There is no doubt that the international aerospace industry will significantly increase its demand for SAF in the very near future due to the mandates and incentives mentioned above. It should be borne in mind that compliance with the blending mandates is the responsibility of the jet fuel producers, who are obliged to pay penalties to their tax authorities in the event of non-compliance. Conversely, airlines are obliged to offset their emissions in the European Emissions Trading Scheme and require emission reduction certificates from their suppliers' mandates. In the US, incentives are provided to both producers (US Blenders Tax Credit) and purchasers of the SAF (RINs).

4.2.1 INCENTIVES AND SUBSIDIES AS PART OF THE INVESTMENT SEARCH

With regard to the stability of the policy framework for investment decisions, it is advisable to include past policy practice in the risk assessment: (1) Will the announced subsidies be available within the announced timeframe, or is there a risk that the start date will be delayed? (2) Will the funding be allocated as announced, or could the qualification process be changed by the funder? (3) Is there any investment protection if the funding provider changes the framework conditions during the construction period of an SAF production facility? (4) Are the so-called grandfather rights granted on the basis of the legal provisions in force at the time of the investment decision or at the time of the building permit, i.e. the continuation of the legal provisions for the operating period of the plant? (5) In the event that applications for support are oversubscribed in relation to the support budget, will the support commitments to the beneficiaries be reduced proportionately, so that the business plan and the expected support amounts will have to be revised? (6) In the case of multiple applications for funding, is there a risk that the cumulation of several funding streams is not legally permissible and may lead to set-off? (7) Are grants paid in tranches and upon achievement of milestones, and what delays in payment can be expected as a matter of commercial prudence?

Government subsidies are typically used to stimulate new business models and investments whose products are in the socio-political or strategic interest of a government in terms of availability and quantity. The reasons for such subsidies are manifold: (1) creation of domestic jobs, (2) better security of supply by avoiding imports, (3) development of new technologies that can subsequently be marketed globally, (4) regional structural support to avoid domestic labour migration and create equal living conditions, and (5) justification of long-term strategies to decarbonise the transport and production sectors.

From an economic point of view, this is a deliberate intervention in the market economy to achieve government goals faster and more sustainably through artificially accelerated development. Government support helps to bridge the so-called "valley of death". This term originates from business research and refers to the transition of research projects into the real world of production and the associated scaling up of technologies. New technologies are supported by venture capital and public research funding until they are ready for production. This phase ends with demonstration or small pilot plants (CAAFI TRL 5), which simulate the production process on a small scale. However, the transition to full-scale production is subject to scaling risk, which is a recurring part of financial planning. Financing the first large-scale plant of a new production process is risky because the measurement data of a demonstration plant cannot simply be extrapolated to industrial production. Past experience shows that such pioneer plants still require process adjustments and sometimes conversions to achieve the desired production result. This delays the start of production and leads to unplanned capital requirements, with a corresponding impact on the break-even point and the achievable profit margin. Projects can fail in this "death zone", which is why project financing for the "first mover" is a challenge. Once the first large-scale plant is in production and producing the expected volumes at the required quality, the "valley of death" is considered to have been overcome and many investors who were not willing to be "first movers" now want to enter the business model as "first followers" (Figure 4.1).

In principle, therefore, government support measures are only temporary, as the formation of a new SAF market should not be supported for longer than necessary. Otherwise, this would lead to a distortion of competition with other SAF manufacturing processes, which would then be at a disadvantage in terms of market participation in the absence of government support. In addition, it has been shown that permanent subsidies do not stabilise a market and that the business model is not self-sustaining once the subsidy ends. Finally, these subsidies tie up considerable financial resources that are not available to the state for other purposes. The problem of the lack of a USP for the SAF production process is repeated, because air transport can continue to operate very cheaply with fossil JET A-1 as long as emissions compensation by offsetting emissions in other sectors outside the transport sector is legally permissible and the certificate price per ton of CO_2 is cheaper than the ton of CO_2 saved by using SAF.

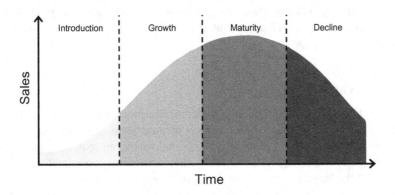

FIGURE 4.1   Illustration of a typical product life cycle (prepared by author).

4.2.2 SELECTING THE RIGHT SAF TECHNOLOGY

As far as raw materials and SAF manufacturing processes are concerned, airlines as end users are agnostic, as long as they do not fear damage to their reputation. As SAF is a commodity with identical technical specifications – but potentially different sustainability scores – there is no customer preference for a particular manufacturing process. Where there is a mandate for the distributor of the jet fuel, any oil company purchasing SAF to meet its quota will select the cheapest supplier whose production and feedstocks are certified through a competitive bidding process. Similarly, airlines that purchase SAF directly will seek to obtain the most attractive price for their tendered volume, for which they will also enter into a longer-term purchase agreement. For the purposes of offsetting SAF in the EU, certification is also required in order to receive the emission reduction certificates directly linked to the physical delivery of the purchased SAF.

In selecting the technology, a decision must be made as to whether the investment funds should be invested in an already tested – and scaled up – SAF production process that requires blending, or whether a process that is in the process of being scaled up should be promoted that achieves the required minimum jet fuel density from its production process and also produces process-related aromatics that are also part of the JET A-1 specification. The risk for such an investment is naturally higher than for SAF's already scaled-up production processes. On the other hand, the new SAF technologies offer synthetic kerosene that can be declared directly as "JET A-1" without further blending with fossil jet fuel if all specification requirements are met. This opens up wider market access, allowing – depending on ASTM regulations – direct sales by the SAF producer to end users without any blending requirement.

The key question of whether the return on investment (ROI) is assured can be answered in the affirmative. Of course, the SAF manufacturer has to be proficient in its technology and professional in the procurement of raw materials. Given the speed with which the transformation of air transport to net-zero emissions neutrality must be driven, demand for SAF will outstrip production capacity over the next 15 years. A tight supply situation with inelastic demand inevitably leads to a market price level at which even the marginal supplier with high production costs can still operate profitably. The distribution of SAF will of course be more complex than it is today, as different distribution strategies will be implemented in parallel in the same market.

4.2.3 LOCATION SEARCH

The search for a suitable production location determines the subsequent transport cost chain. In theory, the production site is located between the raw material production area – usually the "first gathering point" where raw materials are collected from the surrounding agricultural areas and, if necessary, pre-treated – and the location of the largest consumer to be supplied with the synthetic fuels produced. In many cases, the cost of transporting the feedstocks exceeds the cost of transporting the synthetic end products. For this reason, project developers prefer locations that allow short transport distances for the raw materials. Other criteria include the accessibility of the production site, the availability of electricity and water in sufficient quantities and without time restrictions, and the availability of skilled labour to operate the plant. Public subsidies for the establishment of companies and the creation of jobs are further location criteria. It is important to ensure that the chosen location can guarantee an adequate and cost-effective supply of raw materials over the lifetime of the plant. In particular, low-energy raw materials such as forest residues, wood chips or straw make the plant operator dependent on its raw material suppliers, as these raw materials can only be transported over long distances at high cost and with corresponding emission charges in the LCA. In order to reduce such dependencies, preference should be given to locations with good transport links, e.g. close to seaports or railway junctions, which allow raw materials to be delivered from distant growing regions at reasonable cost, if necessary, thus reducing dependency on regional growing regions. Arbitrage pricing, as it is known in economics, can be used as a starting point. In the long term, the price of the raw material will be that of the best available alternative outside the immediate catchment area of the production plant.

For the transport of synthetic fuels to the respective customers or their points of consumption, such as tank terminals (for diesel) or airports (for jet fuel), the transport network and its availability is a central distribution task that must always be included in site planning. For SAF in particular, the use of the existing transport infrastructure for fossil fuels is crucial for cost reasons. With regard to the use of pipelines, it must be clarified whether the operators permit the transit of unblended SAF – as SBC – and what minimum quantities are required for SAF-SBC. However, SAF blends containing FT-SPK or HEFA and meeting the JET A-1 specification can be declared as such and can therefore be used in all pipelines approved for JET A-1. No other SAF blended with Jet A or JET A-1 is permitted at this time. Any future suspension of the current blending requirements as stipulated by ASTM does not mean that such SAF jet fuel is automatically qualified for pipeline transport without further approval.

On the other hand, barges, rail tank cars and road tankers are subject to EI 1530 and EI 1533 regulations for previous cargoes with other products and the specifications for residue-free tank cleaning. In any case, the transport requirements and the choice of transport mode must be determined before the investment decision is made. In Europe, several railway lines are already at their capacity limits, so tank car block trains will have to expect interruptions and stops for train overtakings. The capacity of rail tank cars available for hire is also limited, especially as they can only carry JET A-1 to avoid constant tank cleaning. In the US, bottlenecks in the national pipeline network have occurred repeatedly for various reasons. Special attention must therefore be paid to transport planning when supplying domestic airports in the US. Supply by road tanker is only an option for smaller airports with low throughput. Road transport is relatively expensive and labour intensive. Depending on the location, it is therefore necessary to decide whether it is worthwhile investing in additional tank storage capacity at the production site if this enables larger batches to be transported by pipeline, barge or rail tank car train.

4.2.4 LONG-TERM PURCHASING COMMITMENTS

The financing of further plants for the production of SAF requires a reliable picture of the future demand for SAF, particularly in order to realistically reflect the revenue side. This is a complex task in view of the current situation in a newly emerging market: (1) The volume required for SAF in Europe results from the mandates for the distributors (oil companies) as a lower limit and the voluntary purchases by the airlines – insofar as they wish to purchase additional volumes over and above the mandatory blends imposed on them. In the US, there are no mandates, so tenders are made mainly by airlines and to a lesser extent by oil companies. Willingness to buy is based on the government support available for the production of SAF and the support for the sale of SAF through government instruments (e.g. RINs) in the reference period. The more these subsidy instruments dampen or level the price premium for SAF compared to JET A-1, the greater the airlines' willingness to buy. (2) The competitive situation vis-à-vis other producers of SAF is diffuse. If the production process is identical, the competitors' production costs are fairly predictable, since the main cost difference is derived from the purchase price of the raw materials. In the case of competing SAF production processes, it is much more difficult to assess the production costs because the investment in the plant, the ongoing operating costs and the costs for the raw materials have to be collected or estimated in order to determine one's own competitive situation in terms of costs. (3) In addition to the current demand situation, it is also necessary to forecast the future development of air traffic and its fuel requirements in order to obtain a suitable database.

In order to secure the financing of a new SAF production facility, the project companies negotiate long-term purchase agreements at preferential conditions with the airlines they are interested in before formally deciding on the financing and construction of an SAF production facility, in order to stabilise the cash flow of a new facility once it is in operation. It is understandable that airlines will only enter into a long-term commitment if they can gain a competitive advantage over the competition and if the delivery volume can easily be flown out of one or more airports during the term of the contract. From the SAF producer's point of view, a portion of the SAF production is blocked by secured offtake agreements, but the profit margin from such purchase agreements does not correspond to the long-term profit margin that investors expect from their financial commitment. However, when starting up a new plant, full capacity utilisation is initially more important than the level of return that can be achieved. Once a new producer is established in this new market, it is assumed that the return can be increased to the required level in subsequent years by increasing the sales prices to SAF.

The search for airlines willing to enter into long-term purchase agreements takes between 6 and 9 months, as the relevant departments within the airline have to be involved and the conclusion of a long-term purchase agreement usually requires a decision by the board of directors, as it is a fundamental decision, even if the contract volume could have been decided at management level. From the investor's point of view, the brands of well-known airlines are a confidence-building measure vis-à-vis investors and at the same time a marketing instrument, since smaller airlines like to follow the example of the market leaders. A regional reference to the SAF production site – if it should arise – is very welcome for the participating airline, since every airline cultivates its relationship with the headquarters and the regional population in order to achieve a unique selling proposition (USP) in its core markets. With this in mind, the airline will demand guarantees from the project management that the SAF production facility will be built within the planned timeframe, so as not to suffer any damage to its image as a result of the purchase agreement, which for the above reasons must be concluded before the financing commitment is made. It is therefore common practice to make such off-take agreements subject to the condition precedent of the final financing commitment of all investors involved and to keep the conclusion of the agreement confidential until the condition precedent has lapsed and both parties can announce the conclusion of the agreement simultaneously.

4.2.5 FEED, Permits, Construction Works and Commissioning

Project implementation is a complex task, and sometimes problems arise in the course of implementation that are priced in as "contingencies" or "unforeseen" in the original business plan. Once funding has been secured, the design and detailed engineering planning begins. The common acronym is FEED for Front End Engineering Design. Based on the design documents, the building permit application, environmental impact assessment and other tests for water-polluting substances, fire protection and occupational safety are carried out. At the same time, the site must be checked for contamination and load-bearing capacity. Once all the necessary requirements have been met, the land can be purchased. On the basis of the FEED design, tenders are prepared and suppliers invited for the various construction works and the quantities of building materials are determined. Typically, such a complex structure is awarded to a general contractor to build a turnkey facility. Once all regulatory approvals have been obtained, construction can begin. During the construction phase, the general contractor coordinates the work of the various construction companies, most of which work on the site at the same time. For large projects, it is common for the client to enter into a construction supervision contract with an independent engineering firm. This company works on behalf of the project owner and monitors the progress of the works in relation to the construction schedule and the general contractor's construction diary, and takes its own quality samples (e.g. of concrete) to check the quality of the construction materials used. If there are delays in the construction progress, the construction supervisor and the general contractor discuss what measures need to be taken to make up the delay. If the cause of the delay lies with the general contractor and his subcontractors, the additional costs will be borne by the general contractor. If the delay is due to "force majeure" or is the result of a change in the design requested by the client, the general contractor and the client, with the assistance of the construction supervisor, will negotiate an addendum to the construction contract, the cost of which will increase the construction cost and will be borne by the client. Such supplements are included in advance in the project costing as contingencies for the expected amount. However, if the cost of the supplements exceeds the contingency reserve in the business plan, the project will require additional funding, as typical experience shows that the ability to reduce costs by reducing the scope of works is limited. The additional investment increases the capitalised amount in the balance sheet as well as the depreciation and interest charges in the income statement.

During the construction period, individual sub-sections can be inspected and approved before the plant is completed. Final acceptance of the plant takes place after completion. Any deficiencies identified during this process are rectified within a specified period, either through rework or replacement. If this is not possible, the nonproduction defects can also be compensated by reducing the price of the construction work. The technical acceptance is followed by the legal acceptance and then by the commissioning phase, during which the plant is gradually put into operation and, if it works properly, the products manufactured – provided they already meet all the requirements of the relevant product specification – can be temporarily stored and subsequently sold. If the plant is functioning properly in normal operation, it is released for production.

It is the responsibility of the plant owner to determine when the plant, which is accounted for as an "asset under construction" during the construction period, should be capitalised as a fixed asset and depreciated.

The commissioning phase shows whether the FEED has been planned comprehensively and correctly. The technical design of the plant must take sufficient account of all operating conditions to ensure proper operation. Design errors or insufficiently dimensioned performance parameters become apparent during the commissioning phase and lead to rework to achieve the planned production output. Depending on the complexity of such problems, the start of production will be delayed and additional capital may be required to rectify the problems encountered.

Conclusion: As part of a long-term strategy, investments in SAF production facilities must take into account both the available feedstock and the evolution of jet fuel demand over the lifetime of the facility. In view of the expected market growth (even in mature markets), investments in the production of SAF and renewable diesel are generally advantageous if all the framework conditions are carefully analysed and combined into a coherent concept. The marketability of SAF and especially eSAF is largely ensured by government regulation and/or price-suppressing subsidies. In view of the decarbonisation target, it can be assumed that these framework conditions will remain in place in the long term.

4.2.6 INVESTMENT STRATEGIES

In addition to considering the preferred technology for SAF production facilities, a longer-term investment strategy should be developed that goes beyond the financing of a production facility.

Provided the plant allows for a variable product mix, possibly through the post-treatment of long-chain hydrocarbon molecules, flexible production control can be used to respond to changing demand and market opportunities, improving plant utilisation and profitability. If the production plant and product distribution meet expectations, the financing of further plants of the same type should be considered. If the design is identical, it can be assumed that the technical production risk is low. As long as the SAF market is open for further production volumes, there is basically no sales risk and, ceteris paribus, no price risk as long as all volumes are absorbed by the market. This is not expected to be the case for the entire life of a plant. In this respect, it is necessary to determine the period before the market becomes saturated and price competition increases.

Similarly, the continued availability of government investment subsidies for plant construction and the continued support of the SAF purchase price through increases in mandatory blending rates and/or further tax credit instruments for voluntary SAF purchases need to be examined. In this context, it can be assumed that the JET A-1 price from fossil crude oil will remain well below the price level for SAF in the long term and that the decarbonisation of aviation will only be successful if the substitution of fossil jet fuel is massively advanced. It is therefore unlikely that government support for SAF will be reduced, as this would lead not only to an accelerated increase in CO_2 emissions, but also to a widespread wave of bankruptcies for all companies whose sales strategy for SAF is necessarily dependent on government support or regulation. This assumption applies to all technologies, although production processes that produce marketable intermediates such as ethanol would only be partially affected by such a development. In the absence of alternative measures, the transition of civil aviation to renewable fuels will therefore be irreversible. This also applies to government instruments until the price levels of fossil and renewable jet fuels have converged to the point where further subsidies are no longer needed.

This price convergence can also be achieved through unilateral taxation of fossil fuels. For this to happen, however, the current ICAO agreement on the non-taxation of jet fuel would have to be revoked by mutual agreement. In addition, doubling the price of JET A-1 through unilateral taxation would permanently increase the operating costs of an aircraft by about 30%–40% and lead to significantly higher ticket prices, which would significantly restrict domestic travel in the US, which represents a high social value as a basic public service in the transport sector. Similarly, intra-European travel, especially tourism, would be reduced, with a knock-on effect not only on airline revenues but also on Mediterranean tourist destinations, whose economic product is largely dependent on air tourism. Assuming that these conditions will maintain and increase the use of SAF, further investment in the SAF supply chain is sensible, necessary and profitable in the long term.

Besides CO_2 reduction, scientists are increasingly focusing on non-CO_2 emissions. In addition to the emission of nitrogen oxides, these include in particular the formation of condensation trails ("contrails"), which are formed under certain meteorological conditions from the water content in the exhaust gas stream of engines and contribute significantly to global warming. According to unanimous scientific opinion, the formation of contrails can be significantly reduced by avoiding flight altitudes above 30,000 feet, by appropriate flight routing and by the use of SAF. The latter is mainly due to the better combustion of SAF in the combustion chamber of turbofan engines and the reduced emission of soot particles, which are the nuclei for the formation of ice crystals. From a scientific point of view, it is therefore sensible and advisable to use the highest possible SAF blends in long-haul transport. In practice, however, it is almost impossible to supply long-haul aircraft with high SAF blends, since the large quantities of fuel require aircraft refuelling via hydrant systems, which must be completed within 50–60 minutes in view of the aircraft's flight preparation at the gate.

Supplying long haul aircraft with neat SAF as JET A-1 would be an option if a long haul aircraft could be pre-fuelled with one or two bowser fills of 6,500–13,000 gallons (25,000–50,000 L) of neat SAF before parking at the gate, and the remaining SAF blended jet fuel could then be dispensed at the gate via a hydrant system. With long-haul flights accounting for around 70%–75% of global jet fuel demand, this would make a disproportionately large contribution to emissions reduction, especially as aircraft could then continue to fly the shortest route and at the most fuel-efficient altitudes. This should also be taken into account when deciding on SAF technology.

In anticipation of a medium-term shortage of raw materials for biogenic cultivated biomass and waste materials, the complete decarbonisation of aviation and truck-based transport as a long-term source of demand for diesel fuel will not be achievable. E-fuels will have to be used for further SAF production volumes that are no longer based on biomass. The production of eSAF is based on hydrogen and carbon monoxide as feedstocks. The fuel production of RFNBOs complies with ASTM D 7566 Annex 1, certification for FT-SPK, i.e. the process also used for biomass and MSW gasification. Due to the upstream electrolysis required to produce green hydrogen, significant investment is required in the supply chain, which has a significant impact on production costs, resulting in a cost level that is significantly higher than the production costs of biomass-based SAF technologies. From a commercial point of view, the first step will be to exploit the existing biomass potential and then switch to eSAF production as a complementary supply alternative. As part of a medium- to long-term SAF investment strategy, it is therefore advisable to consider how eSAF can become marketable in competition with SAF. From a regulatory perspective, Germany has decided to introduce an eSAF sub-quota of 0.5% from 2026 and 2.0% from 2030. The agreement in the EU trilogue negotiations (EU Commission, EU Council and EU Parliament) provides for a European sub-quota of 1.2% for eSAF, which can be set higher by national legislators in accordance with EU law. In any case, a sub-quota creates an artificial demand in the amount of the sub-quota, which is demanded by the market in a price-inelastic way. If the regulations announced so far remain in place, this will result in a regulatory demand volume of 220,000 tons of eSAF per year for the German jet fuel market in 2030. If the 1.2% sub-quota is introduced in the other EU Member States, the total demand (including the higher quota in Germany) will be around 900,000 tons of eSAF. As air traffic continues to grow, demand is therefore expected to increase further. As long as the total eSAF production capacity, which must be produced to EU standards in order to count towards the quota in the European Union, does not exceed the minimum quantity required by the quota regulation, it will also be possible to sell eSAF at significantly higher price levels. Timely investment in the eSAF supply chain and its production facilities will ensure that technological development will lead to more efficient facilities and that these can be scaled up to achieve viable production costs that are also competitive with other eSAF producers and ensure an adequate return on invested capital.

Conclusion: Further investment to achieve a pre-defined market share in SAF and eSAF production requires "out of the box" thinking to strategically plan long-term developments. Aircraft have an economic life of 30 years. Air traffic is growing worldwide and other types of aircraft propulsion will not achieve a significant share of traffic until 2050. In this scenario, long-term planning is worthwhile, which can be periodically reviewed and adjusted in terms of objectives and degree of achievement. From a technological point of view, the internal combustion aircraft engine will remain. Supplying the aviation sector with climate-friendly fuels therefore ensures profitable investments in the supply chain.

4.3 SEAMLESS VALUE CHAIN CONSIDERATIONS

As shown in the previous chapters, a significant cost advantage and dominant role of the oil companies has been and still is their control of the entire supply chain "from well to wheel". This makes it difficult for new suppliers to enter the market, as setting up and operating such a supply chain is costly and requires high capital investment. At the same time, this approach allows a reasonable return on investment in all segments of the supply chain, as the proportional allocation of profits between the segments of the supply chain can be controlled through transfer pricing. Depending on the economic requirements of production, segments of the supply chain may be operated in joint ventures with competitors in order to take advantage of economies of scale, although this form of cooperation is not prohibited by competition law. Optimising the just-in-time supply chain also accelerates capital turnover and optimises the use of resources.

In order to replace this successful model in the long term with the new SAF and renewable diesel products, an alternative supply chain for biogenic feedstocks and RFNBOs is needed that is as efficient as the fossil supply chain. It will not be possible to supply the market with approximately 300 million tons of JET A-1 today and 530 million tons in 2050 using small-scale production methods. Small-scale production can serve as a proof of concept to gain insight into the interaction of the various segments of the supply chain, but the approach is not suitable for cost-optimised market supply. Rather, global air transport will require industrial-scale production of SAFs and associated industrial-scale production of the respective biomasses or hydrogen as the main carrier of the RFNBOs. This raises not only the question of the financial viability of the entire process chain, but also the social problem of whether such an expanded production of raw materials is politically feasible. For NGOs and all environmental organisations, the expansion of agricultural land or its conversion for energy production is a favoured issue to use again and again the issues of biodiversity, land use and food competition to prevent the expansion of capacity. In Europe, too, the massive expansion of wind turbines and solar energy is meeting with increasing resistance, as the majority of citizens support the expansion of green energy, but do not want to be personally affected by the resulting construction measures. In densely populated countries in particular, opposition to the expansion of wind and solar energy is likely to grow. While in the US there is sufficient land available for biomass cultivation and wind turbines do not necessarily have to be sited near residential areas, some EU member states will have to import some of their green electricity and cultivated biomass from neighbouring countries due to a lack of land availability. The resulting complexity requires, on the one hand, regulations that are as identical as possible in all EU member states in order to create uninterrupted supply chains. However, it also requires cross-border investments, which require a uniform regulatory framework when the supply chain of SAF production extends across several countries.

Under ideal conditions, an integrated production model for SAF leads to the lowest production costs and thus the smallest price gap between fossil jet fuel and SAF. The smaller the "natural" price differential between the two products for the same use, the smaller the need for government intervention and the faster the transformation of the aviation sector to lower CO_2 emissions. Optimising the production of SAF is therefore not only in the interest of investors in terms of returns, but also in the interest of airlines, who expect a favourable purchase price, as well as in the special interest of environmental policy in the rapid reduction of emissions.

However, the optimisation of the overall process has to overcome trade-offs that stand in the way of an optimal solution to the problem – in the sense of the best possible achievement of the objectives of all individual activities: (1) Determine the factors that can be used to assess whether there is a regional trade-off between food production and energy production. (2) Defining measures to avoid pronounced monocultures in agriculture, but to allow large-scale cultivation of energy crops through appropriate compensatory measures. (3) Rapid development of new waste recycling chains, combined with efficiency analysis of existing regulated waste recycling chains and their continuation or termination. (4) Creation of uniform building codes and zoning regulations for wind and solar energy. (5) Legal assessment of trade-offs between nature conservation and emission reduction measures, and determination of which objectives should be given priority. (6) Accelerate the development of transport infrastructure and electricity transmission infrastructure where this is related to the production of renewable energy. (7) Simplification and acceleration of court and appeal procedures in connection with planning approvals and construction measures for renewable energy. (8) Acceleration of approval procedures, in particular accelerated approval procedures at the EU Commission for state support measures for the market introduction of SAF and renewable diesel.

In order to create a seamless value chain, interest-driven investors need to take the lead in achieving the defined objectives of a business plan. Often, policymakers are not aware of the economic context. It is therefore in the investor's interest to have a sound knowledge base among all stakeholders. Cooperation between the participants in a process chain needs to be institutionalised, as it is highly likely that the process chain will consist of a consortium of different companies. Investors therefore need to determine which elements of the process chain should be realised with their own resources (and financial means) through a make-or-buy analysis, and in which areas they should cooperate with established companies. In any case, it is necessary to have the cooperation of all participants in the consortium and a set of rules accepted by all participants that describes the respective tasks, regulates communication and cooperation in the consortium, and finally, provides for back-up measures in the event of unforeseen performance problems.

1. The most basic form of cooperation is the consortium.

    The members remain legally independent, but there are common rules of conduct and communication for all participants. There are supply and purchase contracts between the participants, but these are coordinated with the production result of the consortium.

2. An extended form of cooperation is the "BGB consortium", which can be used in particular by construction companies involved in building projects. From a legal point of view, this is a civil law partnership (under German law) in which, for example, a bidding consortium can offer an overall price for an entire service. The members of a GbR are legally independent, but a creditor can choose the most solvent partner to satisfy his claims. The paying co-partner will then enforce its payment internally against the liable member.

3. A joint venture is one in which the participating companies have a share in the capital of the joint venture. In this way, the know-how of the participating companies is pooled and the entrepreneurial risk is shared. The exit from a joint venture is subject to conditions, since the purpose of the undertaking can only be achieved through the participation of all the shareholders.

4. The highest degree of integration in a process chain is achieved when the leading company buys up the subcontractors required in the process chain and integrates them into its own process chain or establishes new companies for individual tasks, for which experienced personnel from the relevant industry sector must then be recruited.

The preferred organisational form of an SAF supply chain depends on the achievable economic benefits, the diversification of risks and the long-term nature of the management segment. The secured raw material supply of an SAF production plant is subject to the condition that a raw material price can be achieved that is at least at market price level, but better than market price level under a long-term purchase guarantee. It should be noted that the contractually guaranteed quantity must always be achievable, even in the event of reduced crop yields, which may be at the expense of the possibility of selling excess quantities at better prices.

Integration into a supply chain also depends on the degree of substitutability between services. For example, truck transport with commercially available vehicles can be replaced relatively easily by other providers. In such cases, it should be examined whether regular tendering and reallocation of such services might be more economically successful than a long-term commitment to one carrier.

A closed supply chain for SAF would also include further breeding of energy crops in a holistic approach. For perennial oil crops, such as the Jatropha curcas mentioned above, consistent breeding to maximise oil yield would probably take 15–20 years, with no guarantee of breeding success. From a climate change perspective, timely solutions for immediate emission reductions must be given priority. In this respect, the search for raw materials is increasingly turning to the development of residual and waste materials. Long-term contracts for the disposal of household waste are essential for the construction of waste gasification plants that can utilise the high energy content of household waste in countries without recycling systems, as the construction of the plant is based on the supply of regionally collected waste (Fulcrum, 2022). In this respect, there is a symbiotic relationship between the provision of raw materials and the production of SAF: for the municipalities, the landfill tax, and on the production side, a raw material is used in a cascade that can be procured free of charge or at low cost in constant quantities. Municipalities, counties and states are often shareholders in airport companies and in some cases also owners of the airport fuel depot. In this respect, it is obvious that these owners would also like to build and operate waste gasification plants, for example, in which the locally collected waste can be processed into SAF. In any case, investment projects that can generate additional benefits from the combination of raw materials and processing technology are to be preferred.

This also applies to slaughterhouse waste, animal fats and UCO. In the past, these wastes were "problem wastes" that incurred costs for proper disposal. Today they are products for which a market price has to be paid. Nevertheless, thermal gasification of these wastes is preferable to any other disposal method, as the resulting fuels are considered emission neutral.

For the production of eSAF from hydrogen, the link between hydrogen production and its use as an energy feedstock for power-to-liquid production plants is essential, as intermediate storage of hydrogen is costly and time-consuming. In the overall view of the eSAF supply chain, the generation of electricity from wind and solar energy is at the forefront. According to European regulations, this energy must be used almost simultaneously with the electrolysis of water to produce hydrogen, and according to the "additionality" requirement, electricity generation and electricity use must be closely linked in time; otherwise, there is a risk that green electricity from other sources could be used, which would then be unavailable to existing consumers. Ideally, the hydrogen produced must be processed immediately with carbon monoxide to form a synthesis gas, which is then converted in several process steps to Fischer–Tropsch synthesis and subsequently to naphtha, diesel and SAF. Depending on the technology, both electrolysis and fuel production must run continuously around the clock. During the so-called "dark period", there is no sunlight for the solar panels and no wind to drive the wind turbines. In order to be able to produce 8,400 operating hours per year as continuously as possible, green electricity must therefore be purchased from other sources, but this is sold on site as grey electricity via a "book claim" procedure and its "green" certificates can be used for the nightly use of grey electricity in order to be able to sell all the SAF quantities as green electricity-based eSAF and issue emission reduction certificates for it.

Conclusion: Investment decisions for production facilities are not limited to the core investment. Rather, the entire process chain up to the distribution of the final product has to be considered. Broadening the investment base to include mandatory raw materials and their pre-treatment not only increases investment security but also improves the manufacturing costs of saleable end products and thus their profit margins.

4.4   HOTELLING APPROACH

Hotelling's rule goes back to the economist Harold Hotelling, who in 1929 defined a theoretical pricing mechanism for finite resources that is now considered the central theorem of resource economics: Owners of finite resources are constantly faced with the decision alternative of selling the resource on the market in the short term in order to finance new investments from the yield of the extracted resource (physical capital), the returns of which secure future periods (especially after the finite resource has been consumed) on the profit side. In contrast, there is the option of deferred use. In this case, the use of the finite resource is postponed into the future if the expected future returns from the later provision are higher than the returns from the investments in physical capital made up to that point to reinvest the extraction returns in the event of immediate use of the finite resource.

The two alternatives trigger opposite market reactions. For example, the theory suggests that if the extraction of the finite resource is accelerated, supply will increase and market prices will fall, which will have a negative effect on profits and hence on investment in the reinvestment of the returns. Conversely, if the use of the resource is delayed, the market price for today's production increases, but there is an increased risk that, in the case of later extraction, competing products (such as renewable fuels from biomass or hydrogen), new technologies and changes in consumer behaviour (switching to electric cars) will negatively affect future demand, so that the future market price is lower than necessary to exceed the cumulative returns on the reinvestment of today's earnings.

As a result, the Hotelling rule assumes that the price of the finite resource will increase at the rate of the average rate of return on investment in physical assets, since in this curve the two alternative courses of action are equivalent and form a Pareto optimal equilibrium. Critics of the applicability of the Hotelling approach to the oil market point out that the futures market sometimes shows a price level below the current spot market prices (backwardation). This overlooks the fact that short-term market imbalances – for whatever reason – do not contradict the model's basic assumption of identical marginal productivity of capital and scarce resources (Figure 4.2).

When the environmental model of climate change is added, the demand model changes. For example, the industrialised countries (Kyoto Protocol) are trying to

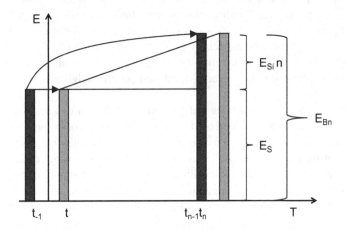

FIGURE 4.2 Chart Hotelling approach. *E*, earnings; *B*, finite resource; *T*, period; *S*, alternative investment; *i*, interest. Explanation: The finite resource (oil) remains unused in period *t*–1 if the deferred use in the future period *tn*–1 is at least equal to the rate of return of the reinvested income in period *t* from the immediate use of the finite resource and the corresponding accumulated interest income (*Ei n*). *EBn* ≥ (*Es* + Esin) holds.

reduce the consumption of oil as a finite resource for climate protection reasons through efficiency improvements, new technologies and alternative energy production. As a result of reduced oil consumption, the finite resource of crude oil will be available for a longer period of time, which means that compared to immediate production and use, a price dampening effect would have to set in due to the decline in demand, especially if the use of mineral oil products is additionally taxed ("CO_2 certificates") in order to create a price umbrella for alternative energies and to induce consumers to reduce consumption by artificially increasing the price of the resource.

The post-Covid19 situation in 2023 with a daily crude oil demand of 100 million barrels per day does not really reflect a reduction in crude oil demand, despite all efforts to substitute oil products with renewables. All published crude oil prices (mainly WTI West Texas Intermediate and North Sea Brent) always reflect daily stock market quotations. There is no inflation-adjusted representation of crude oil prices to verify or refute the Hotelling rule over time. It should be noted that the Pareto optimal equilibrium refers to a static demand situation. With a growing world population and increasing affluence in emerging markets, there is a wealth-induced demand which, in principle, has a price-increasing effect and can overcompensate for the reduced demand in developed countries.

4.5 SUPPLY DOMINANCE APPROACH

In contrast to the Hotelling approach, the "green paradox" (Sinn, 2008) is a model for price-inelastic crude oil supply, which consistently assumes the dominance of demanders over suppliers. Consequently, the effective demand for oil determines the current level of production.

On the other hand, oil-producing countries need a secure and stable revenue stream from oil production in order to meet their current budgetary obligations. The majority of oil-producing countries derive their budgetary resources mainly from oil sales by national oil companies or concessionaires and only to a small extent from domestic tax revenues, especially after the introduction of value-added tax in some Arab countries. It can therefore be assumed that oil production will not fall below a certain minimum level, even if developed countries significantly reduce their demand.

The dampening effect on prices of the expected fall in demand in the developed countries will be accelerated by the additional taxation of oil products (outside the aviation sector). In this situation, consumers in the wealthy emerging markets, which do not have similar excise taxes, will be able to obtain petroleum products more cheaply than in the past. According to the criteria of supply dominance, the underconsumption in the developed countries inevitably leads to overconsumption in the emerging countries as a result of the autonomous expansion of consumption (Figure 4.3).

The intended reduction in the consumption of the finite resource of crude oil, and thus the limitation of additional greenhouse gas emissions into the atmosphere, does not take place under these assumptions (Figure 4.4).

Assuming that the main effect of the Hotelling rule and the "green paradox" is a supply-side quantity constraint, the following assumptions can be made for alternative fuels:

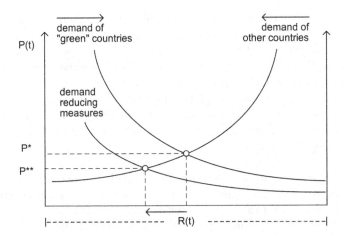

FIGURE 4.3 Principle of the so-called "green paradox" (Sinn, 2008) (prepared by author). Explanation: *P(t)*, price of the scarce resource; *R(t)*, quantity of the scarce resource; *P**, equilibrium price from the crossing point of the demand curves of the "green" countries (from left to right) and the "other" countries (from right to left); *P***, equilibrium price after shifting the "green demand curve" (from left to right) through resource-conserving demand reduction with new intersection of the demand curve of the "other" countries.

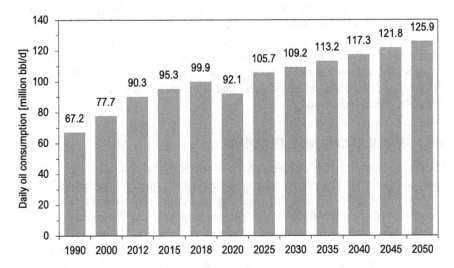

FIGURE 4.4 Daily crude oil consumption 1990–2050 in million barrel/day 2025–2050 estimated figures.

Source: Statista, 2023 based on EIA data (prepared by author).

1. In the long-term trend, the price of crude oil rises in inflation-adjusted terms. As the arbitrage price rises, the production costs of SAF and renewable fuels enter the oligopolistic equilibrium price zone and thus become competitive.
2. The market entry of SAF and renewable fuels does not lead to a global shortfall in demand for crude oil or crude oil products.

3. Economically controlled demand for all types of alternative fuels will develop primarily in countries where governments impose legal restrictions on consumption or artificially increase substitution pressure by taxing fossil fuels, thereby inducing a reduction in consumption and/or investment in low-consumption technologies.
4. Reducing consumption through better energy efficiency does not lead to a global reduction in consumption either. Consumption growth in emerging economies overcompensates for resource and climate protection measures in developed countries.

There is no global substitution strategy for the global balance of increasing oil demand, because conflicts of interest between the countries involved prevent an orderly global transition to an alternative energy supply. As a result, alternative energies will initially be used as niche products in areas where sectoral or regional conditions permit their use. Market penetration will then occur as accelerated resource consumption affects availability and pricing.

Although biofuels are not materially related to fossil fuels, the price of crude oil plus a product-specific processing surcharge for the end products naphtha, gasoline, kerosene, diesel and heating oil remains the global reference for the pricing and competitiveness of SAF and renewable fuels. The advantage of identical product characteristics facilitates the substitution of fossil fuels with all types of alternative fuels. This increases the likelihood that the global price of crude oil will also determine the pricing of SAF, unless there is a significant additional benefit that allows price differentiation or other factors such as tax benefits, regulation and other government intervention determine the market price. Reductions in consumption due to efficiency gains, technological progress and other energy sources are countered by global – cyclical – demand.

4.6 STOCK EXCHANGE TRADING

Market transparency for SAF and eSAF is currently extremely limited. The reasons are simple: (1) The prevailing contract model is a bilateral agreement between SAF producer and SAF user (oil company or airline, respectively). (2) The confidentiality of these contracts limits the publicly available information on pricing and price updates by means of a formula price which, in the case of SAF, is not spot market oriented but reflects the price change of production inputs and energy demand. These variables can, of course, be determined on the basis of stock exchange quotations and their changes. (3) SAF price updates on the basis of exchange quotations for SAF are rarely communicated because the number of reported SAF transactions is still very small overall and provides only a very limited picture of the real market situation in terms of its informative value for market price assessment. (4) Price orientation towards the model of a listed SAF price remains incomplete as long as no free SAF volumes are traded on the exchange. In view of the limited production capacity, available SAF quantities – ready for sale – are only communicated when a buyer withdraws from the contract at short notice and this quantity is then offered publicly. (5) Fuel buyers are therefore faced with the dilemma that they can only derive the market price to a limited extent. In this situation, trust in the seller's honesty is an important element of valuation. (6) Finally, the market price also changes in the context of the events described above and disseminated via the media, which influence the SAF market price for a short period of time. In this respect, a price is always dependent on time and place and may look completely different a few days later.

In the absence of reliable data from the contracts, which are confidential by nature, consultants attempt to develop price trends on the basis of trend developments, comparisons and analogies, whereby the historical price development of similar products since their market launch forms the basis of the trend. Similarly, investment amounts for SAF's production facilities are estimated without the methodology used being presented in a comprehensible manner.

In this respect, the informative value of the publicly available information must be questioned. In particular, it is necessary to examine whether the respective authors are subject to the phenomenon of "dispositional optimism", i.e. whether their perception is based on a fundamentally positive attitude, even if the data situation does not permit such an interpretation (Encyclopedia of Psychology: Optimism, 2023). In medical psychology, it is explained that generalised expectations of results prescribe a positive direction of development, but they leave open whether things will develop by themselves as assumed or whether a personal contribution is required. In the case of predictions about the further market penetration of SAF in civil aviation and the transformation of the emissions situation to "net-zero" in 2050, future developments cannot be scientifically proven on the basis of the literature and are largely incomprehensible because the assumptions that have led to the various assessments cannot be objectively determined. Uncertainty leads to the so-called "illusory truth effect" in perception, whereby a possible – but unconfirmed – development is believed through frequent repetition. With regard to the availability and price development of SAF, the protagonists of the "net-zero 2050" goal mention an SAF share in emissions compensation of between 63% and 70%. The obvious question of why not 100% of CO_2 emissions are compensated by SAF remains unanswered. No distinction is made between the actual emission reduction, since the crediting requirement for SAF in the European ETS is 70% as a minimum (50% as a qualifying requirement in the US), or whether the use of SAF is set at 100% emission reduction, since the basis is SAF production, which achieves a net emission reduction of 70% but is credited at 100%. Finally, a much more realistic estimate of emissions reduction through operational measures in aviation is 3%–5%. Other significant CO_2 savings come from "technical developments" in aircraft and engine technology. It appears that emissions' reductions in the past are extrapolated into the future on a comparable scale without any factual basis. If these uncertainties are used as the basis for the SAF quantity forecast, the question arises as to what criteria could be used to make the SAF price marketable as long as transactions on the exchanges cannot provide a reliable data basis and the tradable SAF quantity does not yet trigger price competition leading to an equilibrium price between supply and demand.

Nevertheless, data validation by reference to stock exchange price data is useful and essential for assessing the price assumptions in a business plan.

In the end, it is irrelevant for the isolated assessment of an investment whether the volume forecasts for 2050 are realistic or whether the uncertainty factor is so large that the range of possible developments does not allow a reliable assessment.

An investment is profitable if the plant produces the planned quantity of product at the planned cost over the estimated lifetime and sells the product at the estimated prices to achieve and maintain the planned profitability of the project. Uncertainties relate to the performance of the plant technology and the sizing of the plant, expectations about the long-term availability of the required raw materials and the assumed raw material prices, the competitive situation with other competitors (with the same or a different SAF technology) and the demand behaviour of the airlines. The demand behaviour of the airlines is completely inelastic with respect to given SAF mandates, as this proportional quantity must be added to the JET A-1 in accordance with legal requirements.

The price of the SAF is decisive as long as the JET A-1 distributors can choose between different offers for the SAF and then decide on the most favourable offer for them. As a cost component, the SAF price is priced into the JET A-1 quantity with its blend ratio and then increases the sales price of the blend. Assuming three times the JET A-1 price for SAF and a blending rate of 5%, the JET A-1 blend price for the airline increases by 10% ($[(0.95 \times 1) + (0.05 \times 3)] = 1.1$). As a mandatory blending rate, this price increase affects equally all airlines that have their aircraft fuelled within the legal mandate. Assuming that jet fuel accounts for 25% of an aircraft's operating costs, the 5% SAF blend will increase aircraft operating costs by 2.5% in absolute terms.

Voluntary SAF volumes will be purchased by airlines directly from the SAF producer. The airline's fuel supplier at the airport must be involved in the process chain to deliver the SAF/JET A-1 blend to the airport fuel depot. The responsible buyer also uses the available stock exchange price quotations as a source of information for the direct tendering of voluntary SAF quantities by the airlines. Voluntary SAF purchases are based on company-specific interests. In the airfreight sector, for example, there is an increasing demand for emissions-neutral transport of goods because the manufacturers of these goods want to sell their customers an emissions-neutral product, which means that both parts and assemblies for assembly must be transported in an emissions-neutral manner, and the distribution of the end products to the market must be emissions-free. Cargo airlines in particular have to meet these requirements in order to continue to be commissioned as carriers. After all, the cost of carbon-neutral transport can be passed on to the customer. Of course, the customers for these transport services also evaluate the stock market data in terms of their willingness to pay a premium for carbon-neutral freight transport. However, airlines can allocate the cost of using SAF blends to specific customers. This is done by first allocating emission reduction certificates from the legally required blending quotas to those customers who request emission-free transport. If this amount is not sufficient, the airline can either purchase additional certificates in the emissions trading scheme or voluntarily purchase additional amounts of SAF, which must then be obtained from a JET A-1 supplier and blended with his JET A-1.

SAF price information from the exchanges is therefore an indispensable element in the pricing policy of SAF producers and serves as a source of information for

SAF buyers. It should be noted that JET A-1 is not traded daily on every exchange, although the exchange information services publish a JET A-1 price for each trading day of an exchange. This is based on telephone interviews in which producers are asked at what price they would sell, and at the same time, buyers are asked at what price they would buy. From this information, the Exchange Information Service calculates a virtual daily price, which is then published. The objectivity of these virtual quotes cannot be verified. It is then binding as a daily value for both seller and buyer within the framework of contractual price formulas.

By including stock exchange trading in the SAF price mechanism, the airlines perpetuate their dependence on speculation-driven price movements. This does not reflect the price as an expression of value or a scarce resource. Rather, the aviation sector is voluntarily paying a speculative return for an everyday product that is regularly in demand in the same quantity, if one ignores seasonal variations in quantity. With the introduction of a SAF market, there is, in principle, an opportunity to emancipate oneself from speculation-driven prices. This is currently the case in the bilateral contracts between SAF producers and SAF purchasers, as the majority of the contracts notified are based on a "cost plus" model, in which the price is based on the cost of production plus a percentage mark-up for profit. These contract models currently depend on individual cases and cannot be generalised. In a free market, each market participant is free to set the price at its own discretion and to update it during the contract period. This is no longer possible if one price model has a dominant position in the market and other contract models are neither offered nor demanded (Figure 4.5).

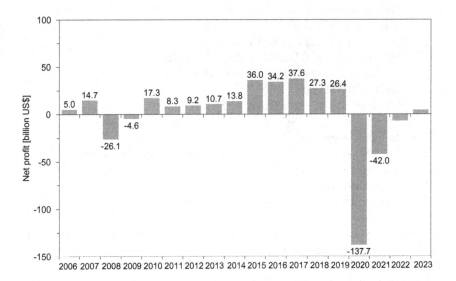

FIGURE 4.5 Net profit of commercial airlines 2006–2021 including forecast 2022 and 2023 in billion USD.

Source: Statista 2023 based on IATA, ICAO and S&P Global Platts (prepared by author).

So far, airlines have shown little inclination to differentiate themselves from their competitors on the cost side. On the contrary, the majority of airlines behave in a price-agnostic manner when it comes to jet fuel as long as a market price affects not only their own operating costs but also those of all their competitors at a comparable level. From a business point of view, this is a surprising behaviour, since the possibilities of reducing one's own operating costs can have a faster and more targeted effect than increasing revenues through additional demand and the use of pricing policy instruments.

As long as the airlines follow the "equal charges principle", this passivity serves the profit margin of the jet fuel producers and SAF manufacturers alike, as they can assume that once a market price has been accepted, it will not be challenged. Especially when pricing has been more or less voluntarily outsourced to stock exchange trading.

Under the current conditions, the SAF price will not be able to catch up with the price level of the fossil fuel JET A-1, because the production costs for the technology and the purchase of raw materials are at a much higher level. In fact, over time, the blending of jet fuel with SAF will lead to a new price development that will even allow oil companies to increase their profits by purchasing SAF and applying unchanged profit margins along the entire process chain. In the above example of a 5% SAF blend (3 times the price of fossil Jet A-1) with a 15% profit margin, a supplier would make the following calculation:

1. Pure Jet fuel
    JET A-1 price: 800 USD/mt
    Profit surcharge: 120 USD/mt (=15%)
    Sales Price: 920 USD/mt.
2. SAF – JET A-1 blend (5% SAF)
    JET A-1 price: 800 USD/mt × 0.95 = 760 USD
    SAF price: 2,400 USD/mt × 0.05 = 120 USD
    Total: 880 USD/mt
    Profit surcharge: 132 USD/mt (=15%)
    Sales price: 1,012 USD/mt
3. Profit increase supplier
    132 USD–120 USD = +12 USD/mt or +10% total increase profit margin per mt of fossil product: 17.36%
4. Price increase airline
    1,012 USD–920 USD = 92 USD/mt or +10%

The beneficiary in this example calculation is the oil company that uses fewer resources from its own production, while at the same time increasing the profit margin on sales from 15.0% to 17.36% relative to its own refinery output.

The reason for this is simple: The acquisition of SAF volumes does not change the production capital used by the oil company. The profit margin is still added to the working capital. By using SAF at a higher purchase price than the production cost of the fossil JET A-1, the total operating costs increase with the effect of the higher profit margin. From the producer's point of view, the coexistence of autonomous price levels for fossil JET A-1 and SAF is therefore entirely desirable. With increasing blending rates, the SAF price level becomes increasingly dominant over time, and with it, a steadily increasing profit margin on the supplier side. If prices rise in the long term (above an average inflation rate), the end-customer business with the airlines will remain profitable.

Conclusion: Over time, price quotations on futures exchanges also develop a normative character for the pricing of SAF products. This trend supports investment decisions as long as no revolutionary technology is able to permanently reduce SAF's production costs to a significantly different level. A lower price basis for SAF will not emerge as long as skimming off the consumer surplus is more advantageous than expanding its own production with lower production costs and gaining market share through price reductions with higher profits than in the skimming variant.

4.7 THE ROLE OF EMISSION REDUCTION CERTIFICATES/TAX CREDITS

4.7.1 EU

The EU Commission, as the executive body of the EU Council, regulates the laws and regulations applicable in the EU Member States through the so-called "directives". The directives represent the minimum standard that is legally binding in all EU27 Member States. However, national governments and parliaments can adopt stricter rules for their territory.

The EU commenced with an emissions trading scheme within the EU Member States since 2005, in which aviation has been participating since 2012. The scope of the EU-ETS (European Emission Trading Scheme) is the trading of "pollution rights", which include the pollutant carbon dioxide (CO_2). An emitter of CO_2 must make a payment for the CO_2 it emits into the atmosphere, which is calculated per ton of CO_2. The emitter must buy a corresponding number of CO_2 allowances for the amount of pollution it has caused. This can be done through auctions, where allowances are sold, or by buying allowances directly from other users who do not need some of their allowances and therefore sell them. As the certificates are dependent on the market, their value fluctuates according to supply and demand.

"Cap and trade" refers to the total quantity of tradable allowances. This is set by the EU and is reduced over time to reduce emissions. The scarcity of supply is intended to drive up the market price of CO_2 allowances and thus motivate companies to engage in climate-friendly production that avoids CO_2 emissions.

For aviation, participation in the EU-ETS currently covers all commercial flights within the territory of the EU27, but excludes flights to supply islands, rescue flights for sick or injured persons and aircraft with a total weight of less than 5,700 kg.

Flights from an EU27 airport to an airport outside the EU27 are not subject to the EU-ETS. The EU-ETS exemption also applies in the opposite direction for flights from third countries to the EU27 (see section 6 for details on the EU-ETS).

The allowance price for one ton of CO_2 was EUR 83.93 on 8 June 2023. For one ton of JET A-1, which produces 3.15 tons of CO_2 when burned, the cost of the allowance is EUR 264.38.

According to the EU Renewable Energy Directive II, the minimum requirement for an SAF to be counted in the EU-ETS is a 70% reduction in emissions compared to the use of the fossil fuel JET A-1. The value of the emission reduction is determined by a so-called LCA, in which the emissions from biomass cultivation to SAF production are measured. This is done either by individual certification of the process chain or by using default values for variables in the LCA. Provided that the SAF produced meets the minimum requirement of a 70% reduction in emissions, 100% of the SAF will count as emission-neutral fuel in emissions trading. The emission certificates are issued by the SAF producer and reported to the relevant national emission trading agency. Due to the price premium of SAF compared to fossil JET A-1, airlines require SAF producers to issue CO_2 certificates for their SAF purchases in order to reduce their compensation for flight-related CO_2 emissions into the atmosphere.

In contrast to the mandatory blending quotas for SAF, where the respective distributor ("seller") of the JET A-1 is obliged to blend, the EU-ETS is an obligation of the end user ("buyer"), who has to make a compensation payment for the pollution caused by the purchase of a fossil jet fuel. In the above example of a JET A-1 price of 920 EUR/mt and a three times higher SAF price for HEFA-SAF of 2,760.00 EUR/mt, the crediting of SAF certificates leads to a price reduction to 2,495.62 EUR/mt. The percentage discount is 9.58%. For comparison: With a HEFA-SAF price of 2,500.00 EUR/mt, the price reduction due to the emission reduction certificates is 10.58%.

The crediting of the certificates in emissions trading is not sufficient to compensate for the price premium between fossil JET A-1 and SAF. Therefore, there is no incentive for European airlines, as the main parties affected by the EU-ETS, to voluntarily purchase additional quantities of SAF beyond the mandatory blending quantities, unless there is a possibility to pass on these quantities, e.g. to a freight shipper who expects an emissions-neutral freight transport and is willing to compensate the airline for the additional costs of SAF.

4.7.2 USA

The US has a fundamentally different approach to the EU: the US Federal Government and the State of California rely on subsidy programmes with financial resources to encourage investment in plant construction and the purchase of SAFs.

At the federal level, the Environmental Protection Agency (EPA) is the central authority for setting fuel standards. The Renewable Fuel Standard (RFS) was first established in 2005 and amended in 2009 (RFS2). During the period of applicability of the RFS2 standard, the production volume for the four categories of (1) cellulosic biofuel, (2) biomass-based diesel, (3) advanced biofuel, and (4) renewable biofuel increased annually to a total of 36 billion US gallons in the final period of 2022. For 2023, 2024 and 2025, the EPA has proposed minimum annual volumes for the four categories, which total 30.43 (2023), 32.80 (2024) and 35.19 billion US gallons of RINs, where one RIN equals one ethanol-equivalent gallon of renewable fuel.

At the state level, California's LCFS, administered by the California Air Resources Board (CARB), is a regulatory limit that will reduce CO_2 emissions from road transport over time. To meet these limits, conventional road diesel must be blended with renewable – synthetic – diesel.

The LCFS requires a continuous reduction in the carbon intensity of road fuels by at least 1% compared to the 2010 baseline, and a reduction of $-20%$ by 2030 compared to 2010. If companies fall below the required carbon intensity in any 1 year, they will be issued with banked credits equal to the amount of the shortfall, which can be used as renewable fuel equivalents in subsequent years to temporarily reduce their renewable fuel purchases. Alternatively, these credits can be sold to other users. In January 2023, the value of a credit per ton of CO_2eq (equivalent) was approximately USD 85, or USD 267.75 per ton of SAF for renewable jet fuel. The credit price fluctuates according to demand and the underlying value to be applied. The LCFS is currently set to remain unchanged with a $-20%$ reduction in carbon intensity in subsequent years beyond 2030. However, there are plans to accelerate the carbon intensity reduction to $-20%$ before 2030.

Conclusion: Due to the specific price situation for SAF and the fact that the product has essentially identical product characteristics to fossil JET A-1, apart from its environmental benefits, there is no USP that offers the end consumer an added value that outweighs the higher price. Therefore, with the exception of artificial interventions by policymakers, the marketability of IT-SAF remains an unsolved problem that is currently only mitigated by legal requirements and subsidies. If the marketability of the product is established with start-up subsidies, a regulated market can develop on this basis. On the other hand, limiting greenhouse gas emissions is a supranational task. The mandatory participation of the European aviation sector in the EU-ETS and the participation of the US aviation sector in the California LCFS are first steps towards using market-based instruments to fairly allocate emissions to their causes and to make them share in the social costs of pollution. (More details on US subsidy schemes will follow in Section 6.4.)

Investment decisions require investment security, at least for the period of depreciation of production assets customary in the industry. In addition, investment decisions are based on a forecast of future market developments, the competitive situation and the positioning of the new investment in the relevant market environment. In a growing market, such decisions are easier to make because demand will remain stable for at least the depreciation period and is likely to increase. This makes it highly likely that the forecast return on investment will be achieved.

The situation is quite different when it comes to financing SAF's production facilities: The demand for jet fuel will increase significantly, but the airlines as the actual consumers of the product show only limited interest in replacing fossil jet fuel with climate-friendly SAF. In addition, the political framework for a product in global demand is diffused from the financial market's point of view. It is difficult to understand why oil companies are not investing in SAF production on a large scale and leaving the technology maturation to third-party financing. Governments in the US and Europe have a very short-term view of the lifespan of SAF plants, and funding measures are also limited in time. The vision of climate-neutral aviation in 2050 is constantly repeated, but there is a lack of congruent action by airlines, today's jet fuel suppliers and politicians to increase the confidence of the capital markets in this technology. The US Federal Government's 2022 IRA can be seen as the first sustainable approach to decarbonising US aviation. Overall, however, further efforts are needed by all stakeholders to mobilise the investment capital required for Net-Zero 2050.

5 Airline Considerations

5.1 MARKET POTENTIAL OF JET FUEL/LONG-TERM JET FUEL FORECAST

A major problem in civil aviation is the lack of meaningful statistics and a globally agreed methodology for forecasting future traffic and its fuel requirements. Future scenarios – if they exist at all – are often driven by vested interests and usually not subject to intersubjective verification. Based on their fleet age and utilisation, airlines determine the age and/or maintenance replacement date for a so-called fleet rollover, where an aircraft type is replaced by a newer model after approximately 30 years of operation. From a business perspective, this is a replacement investment. The productivity of an aircraft is determined by its transport capacity and daily operating time. The data from these parameters are converted into industry standard parameters. There is no reference to future jet fuel demand in such publicly available analyses.

To provide an overall understanding of the future role of sustainable aviation fuels in the context of investment decisions for new aircraft, the cumulative fuel consumption for the operation of an aircraft is calculated below based on publicly available data. This is based on a 30-year service life and a representative rotation pattern for the two aircraft types considered, which is essentially equivalent to the operation of a legacy carrier. The average distance per short-haul flight is 1,220 km (760 statute miles). The cabin layout corresponds to two-class narrow-body seating with an average seat load factor of 85%. For the long-haul aircraft shown, the distance is 9,000 km (5,592 statute miles) and the seat load factor is 90% with a typical three-class layout.

The aircraft used are the A320neo and B737MAX8 for short-haul flights and the A350-1000 and B787-10 for long-haul flights. In both cases, the aircraft performance is very close and can therefore be used as a representative basis for the calculation. The following calculation is not intended to compare the performance of the aircraft models, as the sample routes do not represent the maximum flight distance and maximum possible passenger capacity. Rather, it considers fuel as a necessary operating resource over the lifetime of an aircraft. The figures below are given in kilometres, litres and cubic metres as well as in statute miles and US gallons. Results have been rounded for ease of reading. Aircraft performance and fuel consumption data reflect the author's average calculation of publicly available data. Further details can be found in the Glossary.

5.1.1 CUMULATIVE FUEL CONSUMPTION OVER AIRCRAFT LIFE CYCLE FOR SHORT-RANGE AIRCRAFT AIRBUS A320NEO OR BOEING B737 MAX8

1. Number of daily flights: 7 flights of 1.75 hours block time each and a distance of 1,220 km (758 miles), i.e. Washington (IAD) to Orlando (MCO) = 1,220 km (758 miles) or Frankfurt (FRA) to Madrid (MAD) 1,244 km (773 miles).

DOI: 10.1201/9781003440109-5

2. Ground time between flights: 45–50 minutes for passenger disembarkation, cleaning and passenger boarding, including aircraft refuelling and baggage unloading and loading.
3. Daily aircraft availability from 4.30 to 23.30 (19 hours).
4. Daily aircraft operating hours for take-offs and landings, taking into account night flight restrictions at airports: 6:00–23:00 (17 hours).
5. A gross block time of 12.25 hours is estimated for 7 flights. In addition, there are 8 ground times of 45–50 minutes each. Of these, 6 ground times must be between flights in the 6:00–23:00 time frame. This corresponds to a net operating time of 17.05 hours per day or a gross operating time of 18.65 hours including the ground time before the first flight and after the last landing.
6. The aircraft is available 350 out of 365 days per year. An average of 15 days per year shall be provided for extended maintenance intervals and repairs.
7. The aircraft has an average fuel consumption of 2.8 kg/km or, at standard density of 0.8 kg/L, 3.5 L/km (0.57 USG/mile).
8. Fuel consumption: 3.5 L/km × 1,220 km = 4,300 L (1,135 USG) per flight or 30,100 L per day (7,945 USG/day) and 10,535 cbm per year (2,780,750 USG/year) or 316,000 cbm (83.42 million USG) over 30 years.

This results in the following transport capacity and fuel calculation:

1. The aircraft operates 2,450 flights per year.
2. 1,000 L (265 USG) of kerosene costs USD 620/cbm (USD 2.34/USG).
3. The annual fuel cost is 620 USD/cbm × 10,535 cbm = 6,532,000 USD/year.
4. Fuel expenditure for a 30-year operating life of the aircraft at constant usage and a constant kerosene price of USD 2.34/USG is USD 196 million.

Fuel cost per ticket:

1. The cabin layout of the A320neo/B737 MAX8 is 166 seats. Based on an average seat load factor of 85%, each flight will carry 141 passengers (+5 crew).
2. Fuel cost per passenger is calculated per flight: 4.3 cbm of kerosene × USD 620/cbm = USD 2,666 divided by 141 passengers = USD 18.91/pax/flight. A round-trip ticket will therefore include 37.82 USD for kerosene, regardless of the passenger's booking class.

5.1.2 Cumulative Aircraft Life Cycle Fuel Consumption Long-Haul Aircraft Airbus A350-1000 or Boeing B787-10

1. Number of daily flights: 2 flights of 10.5- and 9.5-hours block time and 9,000 km (5,592 miles) distance, i.e. Los Angeles (LAX) to London (LHR) = 8,780 km (5,456 miles) or Frankfurt (FRA) to Bangkok (BKK) 9,009 km (5,598 miles).

2. Ground time between two flights: 90–150 minutes for passenger disembarkation, cleaning and passenger boarding, including aircraft refuelling and cargo and baggage unloading and loading.
3. Daily aircraft availability 24 hours due to night flights during airport curfew(s).
4. Daily aircraft operating hours 24 hours.
5. A gross block time of 20.0 hours is estimated for 2 flights per day. In addition, there are 2 ground times of 4.0 hours in total.
6. The aircraft is available 350 out of 365 days a year. For longer maintenance intervals and repairs, an average of 15 days per year is provided.
7. The aircraft has an average fuel consumption of 6.03 kg/km or – at standard density 0.8 kg/L–7,54 L/km.
8. Fuel consumption: 7,54 L/km × 9,000 km = 67.900 L (17.980 USG) per flight or 136.000 L per day (35,950 USG/day) and 47,600 cbm per year (12,575,000 USG/year) or 1.43 million cbm (377.2 million USG) over a 30-year period.

This results in the following calculation of transport capacity and fuel:

1. The aircraft operates 700 flights per year.
2. 1,000 L (265 USG) of kerosene costs 620 USD/cbm (2.35 USD/USG).
3. The annual fuel cost is 620 USD/cbm × 47,600 cbm = 29,512,000 USD/year.
4. The fuel expenditure for a 30-year operating life of the aircraft at constant utilisation and a constant kerosene price of 2.35 USD/USG is 885.36 million USD.

Fuel cost per ticket:

1. The cabin layout of the A350-1000/Boeing 787-10 is 331 seats. Based on an average seat load factor of 90%, each flight is operated with 298 passengers (+15 crew).
2. Fuel cost per passenger is calculated per flight: 67.9 cbm of kerosene × USD 620/cbm = USD 42,100 divided by 298 passengers = USD 141.27/pax/flight. A return ticket will therefore include USD 282.54 for kerosene, regardless of the passenger's booking class.

5.1.3 Aircraft Investments versus Cumulative Fuel Consumption

Aircraft manufacturers publish list prices for their aircraft. Depending on the customer and the size of the order, discounts on these list prices are granted, which are agreed between the parties as a confidential agreement in side letters to the respective purchase contract and the content of which is not openly communicated. In addition, each customer can choose from a range of equipment options for its aircraft, which can significantly increase the purchase price. List prices are usually based on a basic configuration, which each customer can modify according to its own requirements. In this respect, published list prices do not reflect the actual purchase price of an aircraft. When compared with published manufacturer discounts, these range from 20% to 50%, depending on the number of aircraft ordered. For the sake of simplicity, a discount of 35% is used below to compare the aircraft investment with the fuel cost.

The estimated net purchase price (after manufacturer discounts) for an A320neo/ B737 MAX 8 is approximately USD 80 million and for an A350-1000/B787-10 is approximately USD 240 million. When comparing the purchase price of an aircraft with the discounted present value of the cumulative fuel costs (net present value based on a discount rate of 3%), the net purchase price of an A320neo/B737 MAX 8 is USD 80 million compared to USD 137.7 million for the JET A-1. This is 1.7 times the purchase price. For the A350-1000/B787-10 with a net purchase price of USD 240 million, the net present value for JET A-1 is USD 622.2 million, or 2.6 times the purchase price of the long-range aircraft.

Assuming an average rate of return on invested capital of 7% for aircraft and 14% for industrial assets, and assuming a 20-year straight-line depreciation for both investment alternatives with a 30-year economic life, the aircraft with an acquisition cost of USD 80 million generates a total return of $(20 \times 0.035) + (10 \times 0.07) = 0.014 \times 80.0 = $ USD 11.2 million. For an investment of USD 80 million in the production of SAF, the airline would earn a total return of $(20 \times 0.07) + (10 \times 0.14) = 0.21 \times 80.0 = $ USD 16.8 million under the same conditions, because the pricing of chemical products is generally set at a higher rate of return than in the transport sector. In this model, the airline would increase its profit by USD 5.6 million and, as a co-owner of the SAF production facility, would most likely be able to obtain a discounted purchase price for its SAF volume, which would further increase its profit through reduced fuel costs.

Understandably, it would not make sense for airlines to invest in fossil fuel refineries because they would have to take on the commercial risk of co-producing multiple products, of which jet fuel is only about 4%–6% of the output, and would also have to invest in the integrated crude oil supply chain. As an alternative product to JET A-1, the new technologies for the production of SAF open up the possibility of integrating the most cost-intensive resource in aviation into one's own value chain in a vertical integration, without taking uncontrollable risks, especially since the production can be technically designed for a very high jet fuel output and the raw material supply is carried out by third parties. The only argument against investing in vertical integration is the pro-cyclical expectation of returns, since a decline in air traffic would also lead to a decline in demand for jet fuel and SAF, and both effects would occur simultaneously. From an economic point of view, a company would solve the "make or buy" situation in a volatile market, where the reduced demand volume in a recession or crisis situation represents the long-term limit of in-house production and the excess volume represents a variable production volume that is purchased on the market according to jet fuel and SAF demand. This scenario is particularly true in the US, where the price premium of SAF over JET A-1 is largely offset by government subsidies.

Airlines operating their hubs in countries with SAF mandates can produce at least the amount of SAF that their suppliers need to meet the government-mandated quota. In such cases, the SAF quantity would be effectively sold to the supplier (including the production certificates) and then bought back (including the certificates) as SAF-JET A-1 blend.

Airline shareholders expect the best possible return on their investment for a defined risk. Analogous to the Boston Consulting Group's portfolio model, in which highly profitable "cash cows" co-finance the next generation of products ("stars"), an airline would need to develop a methodology for its business segments in which investments in related business segments mitigate the company's risk, e.g. through other economic cycles or, in the case of pro-cyclical business development, offer a higher return than the core business.

During the diversification wave of the 1990s, many airlines invested in unrelated businesses without having the necessary expertise. In such cases, the supervisory board's oversight was primarily limited to the regularity of operations and external accounting. In the absence of specific business knowledge in foreign markets, the result was often negative and the invested funds had to be sold at a loss or written off. In principle, investments in SAF production carry a similar risk, with the difference that the technical maintenance staff and the apron staff know the product and regularly check its production route. In addition, the manufactured product is needed for daily flight operations and cannot be reduced or eliminated, as is the case with catering services.

Managing the transition from fossil fuels to renewable fuels is very important for aviation, as the time for fossil fuel jet A-1 is running out. The success of the transition to new fuels depends largely on the behaviour of airlines. With a few exceptions (e.g. British Airways, Cathay Pacific, KLM, United Airlines), the vast majority of airlines remain passive or even hostile, giving the impression that a mandatory blending quota for SAF for all airlines would be detrimental to the aviation industry (Willi Walsh, IATA, 2023). In the absence of a private initiative by the industry association IATA, the climate goals of the Paris Agreement will ultimately have to be pushed through politically. Such decisions have to be made against the resistance and lobbying of airlines – if necessary – to achieve the contribution of the aviation sector to the overall decarbonisation of the transport sector. A voluntary commitment of all IATA member airlines to contribute to the transition from JET A-1 to SAF in a way that is acceptable for the economic success of the sector and does not have a negative impact on flight capacity would push the SAF market into the desired direction. Given that air transport will remain dependent on internal combustion engines for decades to come, and that alternative engines will only be able to provide a limited share of air transport capacity after 2050, switching from JET A-1 to SAF is the only real option for reducing emissions. Regrettably, when this book was edited in September 2023, no such voluntary commitment from IATA could be recognised.

5.1.4   Airline Economics and Airlines' Limited Engagement in SAF

With a few exceptions (e.g. British Airways, Cathay Pacific, KLM, United Airlines), the airlines give the impression that meeting the Paris climate objective is not a task for the aviation industry and are passive or even dismissive in this respect. In the absence of any initiative from the aviation sector to make a voluntary offer to ICAO to come up with a concept that would allow the transition from JET A-1 to SAF to be made with as little disruption as possible and without loss of transport capacity, the governments that signed the Paris Agreement have no choice but to accelerate the sector's emission reductions through legislation or massive financial support programmes. The introduction of mandatory SAF blending quotas is one of several options. However, IATA believes that mandates are harmful to aviation and should be rejected (Willi Walsh, IATA, 2023). Instead, airlines leave it up to passengers to offset the emissions associated with their booking through voluntary offsetting payments as a surcharge on the ticket price when booking a flight. Passengers can choose to use the funds raised to support environmental projects that reduce CO_2 emissions, or to purchase SAFs to be flown on later flights to reduce emissions on those flights. In this way, airlines give the impression that it is the passenger who is responsible for emissions, rather than the airline. This is in contrast to EU legislation, which makes the distributer who puts harmful gases and liquids into circulation (in this case the seller of the fuel) and/or the party that produces emissions from it (the airline, EU-ETS) responsible for the emissions. In both cases, it is not the passenger who is obliged to reduce emissions.

According to publicly available sources, only a small proportion of passengers take up the offer of voluntary offsetting. This also applies to the offer of "green fares", where all emissions of a flight are offset on selected routes with a higher fare (Lufthansa, 2023). As soon as all airlines in Europe are obliged to comply with blending quotas that require an increasing share of SAF over time, the question arises as to why an individual passenger should still pay a surcharge for emissions compensation. Overall, the behaviour of European airlines does not meet the expectations placed on them. On the contrary, they complain about the loss of market share due to distortions of competition, especially vis-à-vis Middle Eastern airlines on Europe-Asia routes. For example, a flight from Europe to the Far East with a flight time of 9–13 hours has to refuel at SAF for the entire route and reimburse the supplier. An airline based in the Middle East, which also offers flights from Europe to the Far East as a transfer connection via its hub in the Arabian Gulf region and where passengers have to change aircraft, only requires the SAF blend on the first flight segment from Europe to its hub; for the subsequent flight segment from the hub to the destination in the Far East, no SAF blend is to be used, which would result in cost savings for the Middle East based airlines compared to the non-stop flights offered by the European legacy carriers.

To understand the cautious or even negative attitude of the airlines, their business model provides arguments worth considering:

1. Airlines generate most of their revenue from the sale of tickets. To do this, they need to offer passengers the widest possible choice of flights to as many destinations as possible so that they will book their flights in the airline's booking system. Airlines therefore consider each passenger's origin and destination. This connection is called "O&D" traffic. The airline has two basic options for making this O&D connection: (1a) The aircraft flies non-stop from the origin airport ("O") to the destination airport ("D"), e.g. from Charlotte, North Carolina (CLT) to Orlando (MCO). On this route, the volume of passengers travelling to Orlando is sufficient to offer non-stop

service between the two airports. At the same time, Orlando is a popular tourist destination where few passengers transfer to other destinations. As a result, the majority of ticket revenue is generated by the CLT-MCO round-trip service. On the other hand, the demand between Mobile, Alabama (MOB) and Orlando is not sufficient to offer a non-stop service. In this case, passengers with different destinations will be combined on one flight to a hub from which several flights will depart to different destinations. (1b) This option is referred to as a "connecting" or "direct" service, although this term is misleading because the aircraft and sometimes the airline must be changed en route. For example, passengers from Mobile bound for Orlando would first fly with other passengers to the Atlanta hub (ATL) and from there transfer to other aircraft for their respective destinations. The Orlando-bound passengers would then fly to Orlando with passengers from other origins. Unlike the non-stop flight, the revenue now has to be spread over two consecutive flights (Figure 5.1).

The bundling of passengers improves the route profitability calculation but does not reach the revenue level of a direct flight. (1c) In a sub-variant of the transfer connection, passengers from different origin airports are flown to a hub, from where they are transported by a long-haul aircraft to another hub, where they again transfer to a flight to the respective destination.

For example, passengers from Hamburg (HAM), Berlin (BER) and Copenhagen (CPH) first fly to the Frankfurt (FRA) hub to connect to a non-stop flight to Singapore (SIN). Some passengers stay in Singapore, but many travellers connect to Bangkok (BKK), Kuala Lumpur (KUL) or Jakarta (CGK). For each "O&D" passenger, the ticket revenue must then be split between three flights if the passenger continues on from Singapore. The revenue situation for each leg deteriorates because the transfer passengers contribute only a margin to the revenue of the route, which does not result in a profit margin for the route, but only covers the variable and part of the fixed costs. The sum of the route revenues thus results from the bundling of different passenger groups to as many O&D destinations as possible. Passengers whose revenues contribute only to cost recovery, but not to the company's profit, are therefore also important for the overall result. The same applies to the calculation of the profitability of each route.

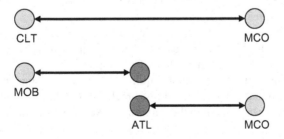

FIGURE 5.1 Operating concept "Point-to-Point" (prepared by author).

2. It follows that airlines operate routes on which they are very profitable, but they also offer routes on which the route result only partially covers the costs. The lower limit of the contribution margin is usually the variable cost of a flight. If these cannot be covered, the airline tries to cooperate with competitors who charge a higher contribution margin for the individual route, but the remaining revenue is sufficient to transport the passenger on a long-haul flight. This is called interlining.

   As a result, airlines operate routes that are not profitable for the reasons described above. With the liberalisation of air transport in the 1980s, traffic rights between countries were liberalised to such an extent that competitors gained access to routes with high demand, and competitive pressure made it more difficult for an airline to unilaterally raise its prices, since its passengers would then switch to competitors with lower prices if they were offered a qualitatively comparable service.

3. This has further implications for airlines' long-haul operations. In order to operate a long-haul route profitably in an airline's network, it is necessary not only to have passengers boarding at the departure airport and disembarking at the destination airport but also to have as many transfer passengers as possible in order to be able to sell the seat capacity offered on the market. Passengers on feeder flights (e.g. from Copenhagen, Hamburg and Berlin to Frankfurt) are carried on these flights without the attributable revenue share covering the variable costs of these flights. A long-haul flight from Frankfurt to Singapore carries a mix of local passengers (from the Frankfurt area) and transfer passengers from different origin airports. Similarly, some passengers remain at their destination in Singapore while others transfer to their destinations there. Revenue shares must also be allocated to the various carriers for connecting flights from Singapore. This complex transportation system is viewed economically in terms of the overall profitability of the O&D connections. The operation of unprofitable routes is therefore accepted if the connection is important for marketing reasons in the portfolio of all destinations, or if the route profitability calculation for the individual route is negative, but the transfer passengers contribute significant revenues for the long-haul flights and the revenue effect of the long-haul route is therefore greater than the loss on the feeder route (Figure 5.2).

4. The operating costs of a route are therefore relevant in the context of SAF blending with JET A-1.

   Looking at the traffic between Western Europe and Asia, there are differences in the operating costs of a route for European airlines compared to airlines operating from hubs in the Gulf region:

   (4a) Longer non-stop flights require a higher crew complement to allow the crew to take the prescribed rest breaks. (4b) A lot of jet fuel is needed for the route. In relation to the maximum take-off weight of an aircraft, the payload weight of the air cargo carried must be reduced in favour of the fuel weight for very long flight routes, resulting in revenue reductions. In addition, the amount of fuel increases by the extra fuel needed to transport

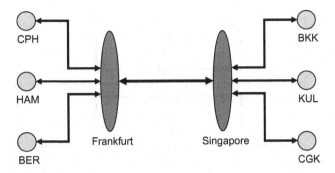

CPH

HAM

BER

Frankfurt

Singapore

BKK

KUL

CGK

FIGURE 5.2 Operating concept "Hub and Spoke" (prepared by author).

the fuel weight over the long flight distance. This additional fuel is called "fuel-to-carry". (4c) For Europe-Far East traffic, the hubs in the Gulf region are not far from the great circle distance of the European and Asian hubs in terms of time, so that no significant flight time increases result from the overall longer flight distance with a stopover in the Gulf region. (4d) If European and Asian airlines were to operate their long-haul flights with a stopover in the Gulf region, no local passengers would fill the vacant seats for disembarking passengers in the absence of corresponding traffic rights, thus enabling corresponding additional revenues. (4e) In this respect, the airlines with their geographically favourably located hubs and the resulting "short" long-haul flights in the Middle East region avoid the additional crew costs, require significantly less "fuel-to-carry" and can adjust the transport capacities in transfer traffic at their hubs by using different long-haul aircraft types on the legs of the journey in line with demand. (4f) While the non-stop long-haul carriers have to refuel SAF blends for the entire route including "fuel-to-carry", the higher jet fuel price for the Gulf carriers will only increase the operating costs of the first flight segment, while no SAF blend has to be refuelled on the onward flight segment to the Far East (Figure 5.3).

5. Regardless of whether the final mandates come into force at the exact level and at the expected time, the competitive situation of European airlines will deteriorate as a result of rising costs per flight and rising costs in route profitability calculations. In the analysis of an O&D connection, revenues initially increase as the route length increases, as long as the additional price for longer routes is accepted by the market. After a certain travel time, however, the O&D connection becomes less attractive, especially if long waiting times between two flights at hubs are unavoidable. In such cases, the fare must be lowered again, despite the increase in distance, in order to generate demand for longer travel times through the attractive ticket price. The attributable costs of an O&D connection react in the same way: As the route length increases, the average cost initially decreases before increasing again due to additional crew work, longer crew rest periods at the destination, additional fuel costs ("fuel-to-carry"),

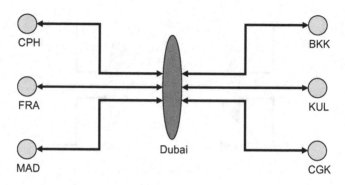

FIGURE 5.3 Operating concept "Mega Hub" (prepared by author).

and a deterioration in aircraft productivity if a long-haul aircraft has to be parked for several hours at the destination in order to pick up connecting passengers on the return flight and land at the origin airport within the airport's operating hours. A significant increase in costs due to the higher SAF price with increasing blending ratios increases the cost pressure and leads to a permanently higher cost level. As a result, the share of profitable O&D connections decreases and the share of loss-making routes increases (Figure 5.4).

5.1.5 SAF MARKET GROWTH AND LONG-TERM ASPIRATIONAL GOALS (LTAG)

In July 2022, the 41st ICAO Assembly adopted the so-called "Long-Term Aspirational Goal" (LTAG) of net-zero aviation emissions by 2050. The subject of Resolution A41-21 is a commitment to the need to meet the +1.5°C temperature target of the UNFCCC's Paris Agreement. No uniform measures are specified, but each State is encouraged to take appropriate action, taking into account its national circumstances and the maturity of its aviation market.

In the "Report on the Feasibility of a Long-Term Aspirational Goal", ICAO has developed three different scenarios as a basis for the development of the SAF market:

1. Fuels Scenario F1 describes the lowest GHG reduction through "SAF production technologies and certification processes are considered that have high achievability and readiness" (ICAO, 2022). Exhaust gas utilisation technologies exist that allow the use of exhaust gases for LTAG SAF production. Low incentives exist for SAF and Low Carbon Aviation Fuels (LCAF) (ICAO, 2022).
2. Fuel scenario F2 reflects the mid-range of possible GHG reductions from SAF. Production technologies and certification processes are in the medium range of achievability and readiness. More technologies are available to enable the use of waste gases for LTAG SAF production, with expanded waste resource volumes. (...) Increased incentives exist for SAF and LCAF production. Carbon Capture Utilisation and Storage is part of the GHG mitigation process (ICAO, 2022).

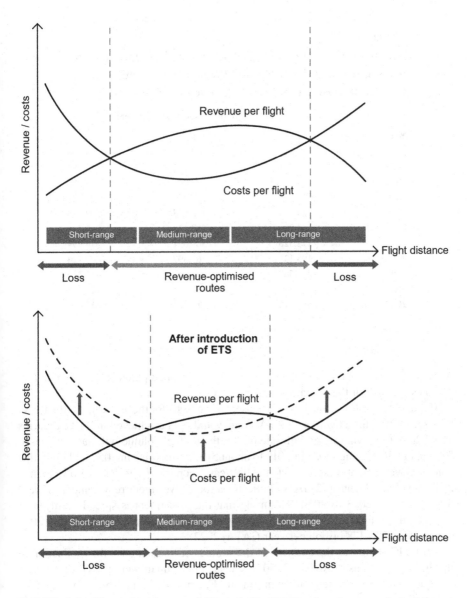

FIGURE 5.4 Illustrative change of Network Profitability based on a substantial cost increase (prepared by author).

3. Fuels scenario F3 reflects the high level of potential GHG reductions from SAF and LCAF. Advanced fuel production technologies and certification processes show low achievability and readiness. However, waste and atmospheric gases are widely used. Sufficient hydrogen production is available, including for hydrogen-powered aircraft. Ground transportation is largely electrified, which will ease the burden on renewable fuels for aviation and shipping (ICAO, 2022) (Table 5.1).

TABLE 5.1

Summary of Unconstrained Projected Fuel Volumes for SAF from Biomass and Solid/Liquid Waste for Selected years from 2020 through 2050 as Calculated by ICAO LTAG

	LTAG-SAF (biomass/waste) [kt/year]		
Year	F1	F2	F3
2020	693	2,821	4,153
2025	1,707	7,571	12,651
2030	4,195	19,965	36,988
2035	10,208	50,113	96,881
2040	24,294	112,828	202,013
2045	54,991	209,138	310,371
2050	112,523	305,031	374,999
2055	195,138	366,897	401,924
2060	277,509	396,494	411,416
2065	334,604	408,635	414,568
2070	365,012	413,293	415,603

Source: ICAO (2021) (prepared by author).

According to ICAO, the Fuel Subgroup experts developed potential sustainable fuel volumes for each of the scenarios shown above.

If the ICAO SAF volume forecast is used as the basis for the assessment, the ICAO "net-zero 2050" target with a 65% SAF rate would not be achieved in Fuel Scenarios 1 and 2! With a dynamic growth rate of 2%, the jet fuel volume will have increased from about 300 million tons in 2019 to about 510 million tons in 2050, although this volume development rather reflects the best-case scenario. In its "Net-zero Roadmap 2050" for Fuel Scenario 2, IATA indicates a total investment requirement of USD 5 trillion from 2023 to 2050. Assuming that the investment is spread evenly over 27 years, the annual investment in SAF production and other technologies after 2035 amounts to USD 178 billion per year (IATA, 2022).

While ICAO leaves the question of the sources of financing for the production of the 305 million tons of SAF in 2050 in scenario FS 2 unanswered, IATA addresses the question of investment financing in the Finance Roadmap, but can only provide indicative order of magnitudes. For example, about one-third of the investment amount is achieved by shifting government support from fossil fuel production to SAF production. The composition of the government support to be shifted from fossil fuels to renewable energy is not specified and should also be questioned, since the energy sector requires an extensive infrastructure and corresponding distribution networks to ensure a country's energy supply. Government spending on the maintenance and expansion of this infrastructure cannot be discontinued without further justification, especially since most of it is also needed for renewable forms of energy. IATA's role as the trade association for the airline industry remains indifferent. However, it is clear that IATA expects all other stakeholders to solve aviation's emissions problems from jet fuel combustion for its member airlines. At least in IATA's publications, with the exception of aircraft weight optimisation and operational measures to reduce fuel consumption, there is no indication of any financial commitment on the part of the airlines to make their own investment contribution to the development of SAF production, for example by stabilizing the critical investments in "first mover" technologies. A notable exception is United Airlines, which has established its own subsidiary, United Ventures, as an investment management company to acquire stakes in SAF production companies.

The open question of how to finance a 5 trillion dollar investment by 2050 therefore remains unresolved for the time being. But it is linked to a follow-up question: At what point will the price of jet fuel rise to the point where demand collapses, making tickets more expensive and private air travel unaffordable for low- and middle-income citizens? Until the introduction of wide-body aircraft, air travel – both business and leisure – was a luxury for people with high incomes. Cost degression due to larger aircraft and the liberalisation of the previously state-regulated air transport market have led to the expansion of air travel, and many tourist destinations generate their gross national product from tourism revenues. Medium-sized companies, which are global technology leaders in their field, need air transport to enable their sales staff to travel to customers around the world. The transformation of air transport to climate-neutral flying is undoubtedly a necessity, in the implementation of which the airlines and their industry organisation must play an important role in order to achieve the best possible success without "stalling". The passivity of many airlines and their lack of commitment to decarbonisation through concrete actions does not send a signal to the financial market that investment in SAF production is worthwhile, and it encourages NGOs and other interest groups to push for their goals, which are eventually taken up and promoted by politicians. While the US is pushing to decarbonise aviation as much as possible by using biomass, the intention of EU regulation is to phase out biomass use in the most controlled way. As a result, SAF products are significantly more competitive in the US than in the EU, as the cost of producing fuels from waste and non-organic products is significantly higher than converting biomass.

Aviation is a global business and aircraft operate in the same atmosphere. It is therefore incomprehensible why the aviation industry does not advocate and enforce globally uniform rules for sustainability measures.

5.2 OLIGOPOLY AS THE PREDOMINANT MARKET MODEL

On the supply side, civil aviation faces an oligopoly of a few international oil companies and a few national oil companies. While jet fuel represents between 20% and 30% of the airlines' aircraft operating costs, depending on the price quoted on the trading exchanges, the distribution of jet fuel represents only about 4%–6% of the oil companies' product portfolio. In addition, the specifications for jet fuel require particularly complex quality control, resulting in additional costs that must be reflected in the product price. As long as the global demand for premium gasoline and diesel remains strong and these products can be sold in many markets, jet fuel will be in constant competition with diesel in terms of profitability, as diesel is straightforward to produce and its sales market is 4–5 times the size of the jet fuel market, depending on the region. Under the given market conditions, several oil companies have been absorbed by the respective market leaders over the past decades or have stopped selling jet fuel. For example, the German jet fuel market, which currently contains about 11 million tons of jet fuel, was the 6th largest sales market for jet fuel in 2019. The oil companies Aral, DEA, ELF, FINA, MOBIL, Q8, TEXACO and VEBA Oil either no longer exist or have stopped selling jet fuel in Germany. This has also reduced the intensity of competition in jet fuel distribution in Germany, which was to be expected given the increasing dominance of spot market quotations as the basis for pricing.

Against this backdrop, the penalties that the EU plans to impose on jet fuel distributors in the event of non-compliance with blending quotas appear in a different context: according to the EU's plans, if the blending quota is not met, not only will a high penalty have to be paid for non-compliance, but the shortfall will also have to be supplied to the market in the following year. If the SAF market in Europe does not develop according to the quota, the companies concerned (distributors) would have to pay penalties even though the amount of approved and certified biomass feedstock in Europe is not sufficient to produce SAF according to the quota. This also applies in the event that SAF production capacity is insufficient to produce the required amount of SAF, or SAF production capacity would be sufficient, but SAF producers sell their production volume to other regions where higher profit margins can be achieved. In such a scenario, some other oil companies may decide to withdraw from jet fuel production. EU regulations require distributors to comply with blending quotas, but they cannot force distributors to build up SAF production through their own investments. If the market does not provide enough SAF to meet the EU's sustainability criteria, legal clarification is needed to determine whether, under these conditions, a penalty payment can be imposed for failure to meet the quota (Table 5.2).

TABLE 5.2
Jet Fuel Consumption per Country in barrels per year/tons per year in 2019

Ranking (2019)	Country	Jet Fuel Consumption [bbl/year]	[tons/year]
1	US	636,333,700	80,935,176
2	China	292,554,800	37,209,996
3	UK	97,071,750	12,346,540
4	Japan	87,578,100	11,139,044
5	Russia	85,629,000	10,891,138
6	Germany	80,683,250	10,262,089
7	Singapore	66,835,150	8,500,752
8	India	64,886,050	8,252,846
9	France	62,327,400	7,927,412
10	UA Emirates	59,768,750	7,601,977

Source: US Energy Information Administration Based on data published by The Global Economy.com (2023) (prepared by author).

Each further reduction of the supplier base strengthens the oligopoly of the oil companies and consolidates pricing by reducing the intensity of competition. While airline purchasing organisations use a portfolio of price negotiation mechanisms – such as reverse auctions, closed bidding and game-theory methods that have proven successful in various negotiation scenarios – fuel purchasing occupies a special position in many airlines that is not linked to other purchasing functions. As a pure commodity purchasing department, there is a tendency for "free-into-plane" supply contracts to be transformed into a contract administration department without a real purchasing function, since price renegotiations are usually already included in the bid price by the suppliers and, in the case of large bid volumes, the suppliers usually prepare themselves to be awarded the contract for only a partial quantity. For this reason, fuel suppliers often only offer partial quantities, which makes it very clear which suppliers need to be contracted in order to contractually secure the total quantity. In this respect, the market entry of SAF producers offers several opportunities for fuel purchasing to break up the existing negotiation rituals and to generate more price movement with innovative negotiation approaches. This is already being practiced in the US through long-term SAF purchase contracts with SAF producers, and blending with JET A-1 is organised in a pragmatic way. The European legal situation allows the purchase of SAF, but the distributor – usually the oil company supplying JET A-1 – is obliged to meet the blending quotas. This situation complicates direct contractual relationships between SAF producers and airlines, but at the same time, it represents a new option that allows the airline as a launch customer to secure advantageous conditions over a longer period of time (Figure 5.5).

5.3 STAFFING AND EXPERTISE OF AIRLINE FUEL PROCUREMENT DEPARTMENTS

A strategic purchasing department has employees with different skill sets. For the conceptual preparation of tenders and new bidding processes, the airline needs managers with an economic or technical background. Mathematicians and physicists provide the basis for the development of new negotiation methods based, for example, on game theory. Such experts can also be provided by specialised consulting firms on a project basis to develop and implement a reset of fuel purchasing. For price negotiations, purchasing needs professionals who have extensive negotiation experience and can therefore conduct professional negotiations with the oil industry's sellers. To monitor the market on commodity futures exchanges, fuel purchasing needs analysts who are familiar with the futures business and can provide a realistic assessment of price trends.

For airlines, aircraft and fuel purchasing are among the internal departments that control and manage large expenditures. It is therefore necessary to entrust these tasks to the best buyers in the company. Purchasers must also have a high degree of personal integrity to act in the best interests of their company and must maintain their independence from all suppliers. The complexity of jet fuel purchasing at US airlines

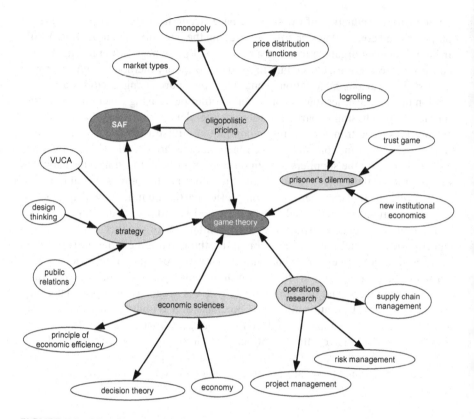

FIGURE 5.5 Visualisation of game theory and the incorporation of SAF based on a chart from Gabler's lexicon of economy (prepared by author).

has always been professional because the logistics of fuel purchasing have always been part of the job. Even 14 years after the first approval of FT-SPK jet fuels, some European airlines still do not have specialised buyers for SAF to take advantage of the expanded range of SAF production processes. These airlines limit themselves due to the lack of competence in elaborating alternative means of supply. Legally, they are not the obligated parties, but passing on all duties to the oil companies leads to limited options. The airlines' cost-cutting programmes, which began even before the Covid19 pandemic, have led to a reduction in personnel in all administrative functions. The aviation initiative "aireg" (Aviation Initiative for Renewable Energy in Germany) was founded in 2011 by the German airlines Air Berlin, Condor, Lufthansa and TUIfly. In 2023, the initiative will have 48 members, but DHL, with its cargo operations, is currently the only airline represented in aireg (aireg 2023). The US initiative CAAFI (Commercial Aviation Alternative Fuels Initiative), with 10 airline members and approximately 600 members in total, receives more attention from the major US airlines, but its activities are guided by its four sponsors, one of which is the national industry association A4A (Airlines for America). The airline members are Alaska Airlines, American Airlines, Atlas Air, Cathay Pacific,

Delta Airlines, Fed EX, IATA, Lufthansa, Southwest Airlines and United Airlines (CAAFI, 2023). Both associations are non-profit organisations that support the promotion of SAF production and the decarbonisation of aviation in a non-competitive manner. In Europe, KLM in the Netherlands and British Airways are both committed to SAF production with the support of their governments.

5.4 AIRLINE PURCHASING COALITIONS

Buyers have a comprehensive toolbox for price-optimised SAF purchasing:

1. Buyers from different companies can form a purchasing pool if they want to buy the same product and achieve a better purchase price by bundling their volumes (OneWorld, 2022). Such a pooling is not objectionable under competition law as long as it does not lead to a dominant position in the overall demand for SAF.
2. In addition, airlines may acquire shares in SAF production companies in order to participate as co-shareholders in the distribution of the company's profits and to obtain a more favourable price.
3. It is also permissible to enter into a combined supply contract for a specified period of time under which the price of fossil JET A-1 is adjusted monthly and a different pricing model is applied to the selected SAF additive.
4. Airlines may also jointly establish SAF production and distribution companies and engage a third party as a service provider.
5. They may trade in SAF, but must comply with the tax framework in the case of a physical sale of already produced quantities.
6. They may trade tax credits (Renewable Identification Numbers, RINs) and buy and sell CO_2 reduction certificates.
7. They can arrange swaps with friendly airlines to meet their sustainability requirements and avoid losing subsidies or credits abroad. For example, airline A in the US can transfer a partial amount of US SAFs to airline B. In return, airline B in Europe buys a correspondingly larger amount of EU SAFs and transfers an identical amount to airline B when the aircraft refuels at a European airport. The two airlines agree bilaterally on quantity control and price compensation (Figure 5.6).

5.5 DECARBONISATION OF AIR TRAVEL: OFFSETTING AND REPLACEMENT OF FOSSIL FUELS BY SAF

The options available to airlines for reducing CO_2 emissions must be differentiated according to the operational concept and the aircraft used.

1. By definition, the classic legacy carrier operates a comprehensive network of short- and long-haul routes and bundles transfer options via hub airports. The business objective is to offer a comprehensive range of flights designed specifically for business travel. This requires daily flights to destinations

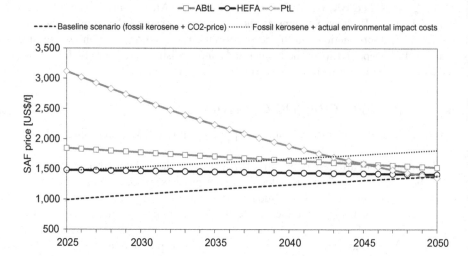

FIGURE 5.6 SAF price development until 2050 (unblended).

Source: PWC Strategy & Analysis (2022): The real cost of green aviation. Evaluation of SAF ramp-up scenarios and cost implications for European aviation sector (prepared by author). ———: HEFA (average), ----: ABtL (average), – – –: PtL (average). Explanation: PWC forecasts price parity of HEFA and fossil kerosene already ahead of 2030, while PtL will reach price parity between 2040 and 2045 comparing SAF products with fossil jet fuel including CO_2 pricing. According to PWC, SAF will contribute to net-zero 2050 only with 53%. This percentage reflects the lowest SAF share projection identified in all publications as found in the bibliography.

and, on high-demand routes, a correspondingly high frequency of flights in order to offer travellers a flight that fits their schedule with as many staggered departures as possible. Flights up to 1,500 km (930 miles) are defined as short-haul. Flights longer than 1,500 km are considered medium- and long-haul. On long-haul wide-body aircraft, the available payload reserve is regularly used to transport air cargo in the belly holds. With the additional cargo revenue generated, the legacy carrier improves its operating results. The take-off weight of long-haul aircraft is therefore regularly at the limit of the maximum take-off weight. Fuel consumption during the climb to cruising altitude is correspondingly high. Jet-powered aircraft can reach high airspeeds and are therefore certified to a maximum cruising altitude of 42,000 feet. Depending on their weight, these aircraft can only cruise to about 30,000 feet after take-off. As flight time and fuel consumption increase, the aircraft becomes lighter and can gradually climb to higher altitudes, where the lower density of the air reduces drag and improves fuel efficiency. The higher the altitude, the more economically the aircraft flies. However, the aircraft's emission balance worsens because CO_2 has a residence time of 200 or more years in the tropopause due to the low air circulation (DLR, 2007).

2. The so-called low-cost carriers operate according to a cost-optional operating concept: in order to be able to offer the passenger flight service at

a low price, the aircraft must be utilised to the maximum every day and as many passengers as possible must be transported in order to achieve a seat load factor of more than 95%. Low-cost carriers mainly fly short-haul flights of 1.0–1.5 hours in order to operate up to 10 flights per day. They use single-aisle aircraft from the A319–A321 family and Boeing B737-500–B737-MAX 10, and they do not carry cargo in order to keep ground times between flights as short as possible. The pricing model of low-cost carriers consists of low fares. Ancillary revenues increase the revenue per passenger for seat selection, in-flight food and beverages, credit card reservations and baggage check-in fees. Flight time at cruising altitude is limited due to short-haul connections. CO_2 emissions per passenger are more favourable than for legacy carriers due to the short flight time and narrow cabin seating to maximise the number of passengers allowed. In addition, low-cost carriers operate modern fleets to minimise flight delays and cancellations due to the technical reliability of newer aircraft.

3. A subset of low-cost carriers are leisure airlines, which primarily fly tourists to their vacation destinations. These airlines also fly according to a published schedule that can be booked by anyone, but they carry groups of travellers booked as a contingent by tour operators. Passengers always check in their baggage, which is then transported in the cargo hold. In addition, the flight duration is longer than that of a traditional low-cost carrier, at 2.0–4.5 hours. In Europe, vacationers are flown to destinations around the Mediterranean and the Canary Islands. In the US, Florida, the Caribbean and the Hawaiian Islands are at the centre of the leisure segment. European leisure airlines operate at full capacity during the peak season from April to September. In the fall and winter months, the number of flights is greatly reduced and various aircraft are taken out of service for several weeks at a time. The carbon footprint of leisure airlines is worse than that of traditional low-cost carriers because the fleet is older and the flight duration is often medium range, between 1,500 and 4,000 km (930 and 2,500 miles). These flights are usually non-stop, as passenger transfers are time-consuming and limit the use of the aircraft. Because of the flight times, night flights can also be operated if the destination airport is open 24 hours a day.

4. As a subset of long-haul traffic, Arabian Gulf airlines operate a single-hub concept with mostly short long-haul routes connecting airports with sufficiently large catchment areas to their hub. Flight times are between 6 and 8 hours, allowing for high frequencies as their hubs can be served without night flight restrictions. Flight routes are based on the great circle distance, which optimises flight time and fuel consumption. Flight preparation time for a long-haul flight is approximately 2 hours. Therefore, passenger transfer time does not adversely affect aircraft utilisation. The CO_2 consumption per passenger is more favourable than in the long-haul operations of the legacy carrier due to the route length and the use of modern fleets.

The EU27 regulations on CO_2 offsetting and mandatory blending are detailed below:

1. Legal requirements for blending always relate to the distributor of the fossil A-1 jet fuel. In most cases, this will be the manufacturer of the JET A-1, who either produces SAF himself, buys SAF from other manufacturers and blends it into his product, or blends SAF owned by an airline and then delivers the production batch to the airline, charging only for the fossil portion of the batch, since the blended SAF was already owned by the airline.
2. The distributor is not required to produce each production batch in the exact blending ratio required by law. Rather, quota compliance applies only in aggregate to all quantities produced in a given period. The distributor may, at his discretion, provide individual batches with higher blending ratios than those prescribed, or produce batches with lower blending ratios or none at all.
3. With regard to the price commitment, the distributor shall prepare a mixed calculation based on the fossil quantity and its unit price as well as the SAF share with the SAF-specific price.
4. The distributor shall receive the emission reduction certificates from the SAF manufacturer for the quantity placed on the market by the distributor. Usually, fuel supply contracts contain a provision according to which the airline receives the same amount of emission reduction certificates for the delivered amount of SAF.
5. It is unclear what regime would apply if the distributor is unable to purchase on the market the amount of SAF it needs to fulfil its mandate obligation. Alternatively, the distributor could purchase quantities of SAF that are eligible in the US under US rules but do not meet EU requirements. The alternative would be to inform the relevant tax authority that such SAF are not available. From a legal perspective, purchasing unconventional SAF would be a mitigating measure, but it is questionable whether this approach would be recognised by the authorities and the penalty reduced accordingly. If the purchase is not possible, the question arises as to whether the legal regulation remains suspended until the delivery can be made.
6. If there is an increase in penalty payments in the EU27 even though it is impossible to procure the necessary quantities, some oil companies may refrain from producing JET A-1 if the sum of penalty payments is higher than the profit margin generated by the distribution of JET A-1. There is no legal obligation for oil companies to produce SAF in their own plants, and the sellers of SAF will sell their quantities to the buyers who pay the highest remuneration.

While the distributors have to comply with the quota regulations, the airlines are obliged to participate in the European Emissions Trading Scheme as "status polluters", i.e. according to the German Emission Protection Act. The EU-ETS stipulates that an emitter must use a quantity of CO_2 certificates equal to its CO_2 emissions to compensate for the environmental damage. JET A-1 produces a total of 3.15 tons of CO_2 per ton of kerosene when burned. In 2012, intra-European air traffic was included in the emissions trading system, which only applies to flights within Europe. However, in previous allocation periods, airlines were given large quotas of free CO_2 allowances to prevent the financial burden on airlines from skyrocketing. However, the amount of free CO_2 allowances has been continuously reduced and will reach 0% by 31 December 2025:

1. Airlines (still) receive free CO_2 allowances, but the amount is already decreasing by 2.2% per year and will be reduced by 25% in 2024 and 2025 in order to reach the "fit for 55" target of a 55% reduction in emissions in 2030 compared to 1990.
2. Airlines must purchase or auction the missing allowances for their intra-European flights.
3. If airlines purchase SAF volumes – whether mandatory volumes under the quota system or voluntarily purchased volumes – they can use these allowances and purchase correspondingly fewer allowances. From 1 January 2026, all EU-ETS allowances will be subject to a fee.

To better understand which flights are subject to the EU-ETS and the EU mandatory neutrality obligation, some sample flight routes are presented below:

1. A US airline operates a flight from Atlanta (ATL) to Stuttgart, Germany (STR). The flight departs from an airport outside the EU and is exempt from all emissions offsets as an international flight. However, Stuttgart is only a stopover with disembarking passengers. The flight then continues under the same flight number from Stuttgart to Prague, Czech Republic (PRG). The continuation of the flight takes place within the EU27 member states. This flight segment is therefore subject to the EU-ETS and the CO_2 emissions must be offset by the airline. In order to comply with the blending quota from 2025 onwards, the fuel supplier at Stuttgart Airport will deliver the SAF-JET A-1 blend for aircraft refuelling without the need for a separate order from the airline. The delivery price will increase accordingly to reflect the SAF component.

   The US airline starts its return flight at Prague Airport and continues from there to Stuttgart. Since this is an intra-European flight, the airline has to compensate the CO_2 emissions according to the EU-ETS. For the refuelling in Stuttgart and Prague, the US airline will receive emission reduction certificates for the proportional SAF blend – as soon as the blending regulations come into force – and can use these to reduce the purchase of chargeable emission allowances. This also affects the refuelling in Stuttgart for the return leg from Stuttgart to Atlanta. This is an international flight that is not subject to emission charges, but the amount of fuel used includes the SAF blend because the refuelling takes place at an airport within the EU27. The airline can use the corresponding emission reduction certificates to reduce its EU-ETS obligation.

Of the four flight segments of the ATL-STR-PRG-STR-ATL rotation, two flights will be subject to EU-ETS offsetting and three aircraft fuelling operations will receive a 2% SAF blend from 2025 onwards.

2. A Dutch airline operates a flight from Amsterdam, Netherlands (AMS) to Princess Juliana Airport in St. Maarten, Leeward Islands (SXM). The southern part of the island is legally part of the Kingdom of the Netherlands and thus part of the EU. They are "Overseas Countries and Territories of the EU" where EU law applies. Consequently, the flights from Amsterdam to St. Maarten and back to Amsterdam are intra-European flights. As a result, these flights are subject to the EU-ETS, and from 2025 onwards, aircraft refuelling at Princess Juliana Airport will also have to use a 2% SAF blend – at least virtually.

3. A French airline operates a flight from Paris, France (CDG) to Fort-de-France, Martinique (FDF). Martinique is also part of the Leeward Islands, but it is French territory and one of the so-called "outermost regions" (of the European Union). It is also subject to the EU-ETS, including the obligation to offset CO_2 emissions in the EU-ETS for outbound and return flights, as well as the EU blending obligation from 2025. Provided that the fuel suppliers in St. Maarten and Martinique are based in Europe or have branches in Western Europe, no SAF needs to be physically transported to St. Maarten or Martinique. The EU requirements are met if the JET A-1 quantity provided at these airports is increased by the corresponding SAF blending at another European airport and the allowances are allocated to the virtual SAF quantity at the respective airport in the Caribbean.

According to the EU Regulation on public service obligations for the outermost regions, an EU-ETS levy does not apply if the traffic volume in such a region is (1) less than 30,000 seats per year or (2) less than 243 flights per period or 729 flights in absolute terms over three consecutive four-month periods. This de minimis rule in the EU also applies to all of an airline's flights if (3) the total CO_2 emissions are less than 10,000 tons of CO_2 per year or (4) the aircraft used have a take-off weight of less than 5,700 kg.

4. A Scandinavian airline flies from Copenhagen, Denmark (CPH) to San Juan, Puerto Rico (SJU) as a stopover. From there, the aircraft continues under the same flight number to Martinque (FDF). From there, the aircraft returns non-stop to Copenhagen, Denmark (CPH). Such a routing is called a triangular flight. The CPH-SJU flight is an international flight and therefore does not count towards the EU-ETS. The onward flight from SJU to FDF is also an international flight and therefore not included in the EU-ETS. The return flight from Martinique is also an international flight, but due to the classification of Martinique as a French overseas department, it is considered an intra-European flight for which allowances must be purchased. As of 2025, aircraft refuelling will take place in Copenhagen and (virtually) in Fort-de-France with a 2% SAF blend. The corresponding allowances can be used by the Scandinavian airline to reduce EU-ETS charges.

5. A German airline operates a non-stop flight from Frankfurt (FRA) to Los Angeles (LAX) and back to Frankfurt. The outbound and return flights are international and therefore exempt from the EU-ETS, but will be covered by CORSIA in the future.

In Los Angeles, the airline voluntarily fills a SAF-JET A-1 blend with 10% SAF from US production. It receives proof from the SAF producer that the product meets the sustainability requirements of the Environmental Protection Agency (EPA) in the US and achieves an emission reduction of at least 50% compared to fossil JET A-1. In addition, the airline will receive RINs from the SAF producer corresponding to the SAF quantity. The airline can use the RINs in the US because they are tax credits that can be used to reduce an existing US tax liability. However, the airline cannot count the amount of SAF fuelled (within the fuelled blend of SAF and JET A-1) in the EU-ETS because the EU sustainability criteria are defined differently than in the US. Thus, the SAF purchased in the US would have to achieve at least a 70% reduction in emissions compared to JET A-1. In addition, the origin of the SAF would have to be certified in the US according to European criteria in order to be recognised in Europe. Once the LAX fuelled SAF volume counts as a CORSIA eligible fuel, the airline can offset related CORSIA obligations by using the certificates issued by the US jet fuel supplier.

6. A French airline will fly from Paris (CDG) to Berlin (BER) and back to Paris in 2026.

As intra-European flights, both flights are subject to the EU-ETS. In 2026, there is a blending obligation within the EU27 of 2% SAF to JET A-1. In French refineries, a selective SAF is blended with fossil JET A-1 according to ASTM D 7566. In Germany, there is an obligation to meet a PtL sub-quota of 0.5% within the SAF quota in 2026. German jet fuel producers are therefore required to use 1.5% SAF from biomass-based production processes and an additional 0.5% SAF from power-to-liquid kerosene, which is significantly more costly to produce and sell in Germany than the other production pathways. It also applies to this example that the physical supply of an aircraft at an airport is independent of the availability of blended fuel. The corresponding blending quotas are included in the price calculation by the suppliers and must be compensated by the airlines. The fact that SAF is only distributed at a few airports and therefore with higher blending rates is irrelevant from the point of view of customs and tax authorities as long as the total amount of SAF (and the sub-quota of PtL in Germany) has been demonstrably produced and distributed in the course of a year.

The above sample flight connections illustrate the complexity of the EU-ETS and the blending obligations of jet fuel distributors to be implemented in national law.

There appears to be a regulatory gap in the EU legislation: Jet fuel distributors – typically the oil companies that produce fossil jet fuel – are required to comply with the EU blending quota, but not to produce SAF! As a result, the majority of today's jet fuel producers limit themselves to purchasing SAF from the market, i.e. buying a certified product to add to their fossil A-1 jet fuel. If a distributor does not fully comply with his blending obligation, he faces very high penalties and the obligation to place the missing amount of SAF on the market in the following year in addition to the current year's amount. If the market does not provide sufficient quantities of EU-certified SAF, a distributor will be able to argue that it was unable to obtain sufficient quantities of SAF for its production on the market due to a lack of sufficient supply, regardless of the price set by the SAF producers.

From the airlines' point of view, it will be an impossible task to increase the SAF content significantly above 30% as long as the current ASTM requirements for SBCs remain unchanged.

1. The SAF blend must be designed to meet the lower density limit for JET A-1. Due to technology processes, only one of the seven approved technology pathways currently achieves a product density within the JET A-1 specification.
2. Similarly, JET A-1 requires an average aromatic content of 17% to ensure the tightness of rubber seals in the aircraft fuelling system. (Van Dyk, 2022).

    With this in mind, ASTM has only approved the SAF technology pathways as a synthetic blend component. Notwithstanding this, individual airlines and their industry association, IATA, are projecting that they will meet approximately 70% of their fuel requirements with SAF by 2050. Under current conditions, this statement is not feasible.

    Since the blending mandates of the governments affect the distributors, the airlines can follow this development without any activities of their own. Once again, however, the EU and its member states are failing to provide any feedback between the legislative setting of emission reduction requirements and their technical feasibility.

Airframe and engine manufacturers state that all new aircraft and engines will be able to fly with an aromatic-free neat SAF by 2030, but this statement is based on a number of test flights in which the existing O-rings have been replaced with new products to eliminate the risk of fuel leakage. To date, the issue of lower fuel density has not been resolved, which means that a route flown with neat aromatic-free SAF requires a greater amount of fuel than a flight flown with JET A-1, as the lower energy density of the fuel results in a higher volume of fuel consumed. However, for aircraft that are limited by fuel weight, but not volume, the fuel efficiency is actually improved due to the higher specific energy of aromatic-free SAF. Similarly, Boeing had already begun test flights in 2015 with a synthetic diesel blend for JET A-1. Boeing's intention was to accelerate the SAF ramp-up. However, at the time of this writing, the synthetic diesel blend with JET A-1 has not been qualified by ASTM.

This leads to the following conclusions for airlines:

1. As long as ASTM, as an independent certification organisation, acts conservatively in the interest of aviation safety and only allows changes in the use of SAF when it is proven by practical tests and evidence in flight operations

that higher blends are not detrimental to safety, announcements that give the public an inaccurate picture of the development should be avoided.

2. Even if aircraft delivered from 2030 onwards no longer rely on the presence of aromatics in jet fuel, there are still approximately 25,000 existing aircraft whose fuel systems rely on the presence of aromatics. It will be logistically impossible to provide two different types of Jet A-1 at airports. In addition, the risk of misfuelling increases when an older aircraft is refuelled with an aromatics-free jet fuel.

3. Fuelling exclusively with clean SAF requires a complete overhaul of current fuelling procedures, where fuel is sold and delivered to aircraft by volume rather than by weight. It is technically possible to measure the current density in the flow at the tanker during aircraft refuelling and to increase the flow rate accordingly until the required energy content is delivered. Such a procedure would first require reprogramming of the aircraft's on-board computers and also investment in the conversion of the tanker's measurement equipment, including electronic data transmission to the cockpit of the aircraft being refuelled, to the airport fuel depot operator and to the sales company on whose behalf the aircraft is being refuelled. Such a procedure must be applicable worldwide, from major international airports to regional airports in developing countries.

4. This would require a global shift in jet fuel sales prices from volume-based to weight- or energy-content-based prices.

5. Consequently, airlines should not announce higher blending rates until they have contractually secured access to corresponding SAF volumes that contain sufficient levels of aromatics such that the above issues do not apply. The already approved technology pathway "Catalytic Hydrothermolysis Jet fuel" (CHJ) basically fulfils these requirements and would be able to replace JET A-1 as a pure SAF in large-scale production. However, the technology has not yet reached the required Technology Readiness Level (TRL) 8. Like HEFA, CHJ uses vegetable oils, animal waste and used cooking oils as its raw material base. As such, the limitations on feedstock availability also apply to this technology.

5.6   AIRLINE OPTIONS TO STIMULATE THE SAF SUPPLY CHAIN

As a first step towards further consideration and action, each airline should first collect and model the current emission levels of its own flight operations and ground facilities. The actual data can then be used in further steps to extrapolate values for the coming years. An essential part of the data collection is the current aircraft fleet as well as its changes due to the retirement of older aircraft, the commissioning of new aircraft as replacement investments and the commissioning of new aircraft as expansion investments to secure existing market shares in individual traffic regions or to gain further market shares in strategically important markets. The main determining factors are the specific jet fuel consumption of an aircraft type and the number of TKOs (ton-kilometres offered) provided by this sub-fleet. With the gradual renewal of the aircraft fleet, the first effects on the development of consumption, ceteris paribus, can be determined. It is possible and probably expected that the range potential of new aircraft types will be exploited, which could change the route length and the operational profile of this sub-fleet. In such a case, fuel savings from fuel efficient engines would be offset by additional emissions from a changed flight profile. In any case, the first step should be to take stock and, based on that, to examine the likely evolution of the fleet and emissions at least up to the year 2040.

The following graph (illustrative) shows that aviation emissions increased in absolute terms up to 2019. This shows that all emission reduction measures, in particular the reduction of aircraft specific fuel consumption, have not led to a reduction in emissions. On the contrary, global market growth has been so strong that it has overcompensated for all emissions reductions. The decrease in emissions shown in 2020 is based solely on the sharp drop in the number of aircraft movements due to the Covid19 pandemic. From 2022 onwards, civil aviation is on a recovery path and will return to the emissions level of 2019 by 2024 at the latest, unless airlines take appropriate measures to avoid exceeding the previous peak (Figure 5.7).

Each airline should therefore first present its own historical emissions curve and extrapolate its future development in the form of scenarios based on fleet development and the flight plan. At the same time, the net-zero 2050 target should be visualised to determine the airline's need for action. Exceeding the 2019 emissions level in the years from 2024 onwards is a possible, albeit undesirable, situation.

The airline's actions should be designed for sustainability and effectiveness. The Marginal Abatement Cost Curve (MACC) is one way to visualise the options.

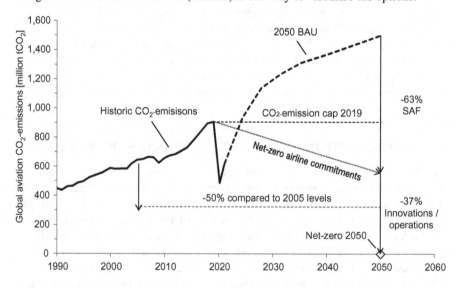

FIGURE 5.7 Illustration of aviation emission reduction requirements Commencing 1990 as base year and 2005 as reference year resulting in a SAF net zero commitment of 63%–65% SAF as emission reduction (prepared by author).

5.6.1   MACC – MARGINAL ABATEMENT COST CURVE

The concept of Marginal Abatement Cost Calculation describes the marginal cost of reducing pollution by a unit of quantity. In the case of aviation, for example, this would be the cost of reducing emissions by 1 ton of CO_2.

Even before the Covid19 pandemic, airlines measured the amount of unused water in lavatories on long- and short-haul flights to determine the actual amount of fresh water needed to reduce emissions. The amount of fresh water carried was then adjusted to the expected demand. This measure reduces the weight of the aircraft, resulting in lower fuel consumption. In addition, reducing the amount of fresh water also reduces the cost of fresh water. In this case, there are "negative costs" because maintaining the previous practice of filling the tanks to the maximum amount of fresh water caused higher water charges than the emission-reducing innovation of reducing the amount of water. All measures that cause negative marginal costs are positive for the environment and the airline's operating results. Other measures incur costs but reduce emissions. A practical example is the substitution of the fossil fuel JET-1 by SAF. Since SAF is more expensive than JET-1, the airline incurs additional costs. At the same time, the airline emits less CO_2, so it has to offset less CO_2 on an intra-European flight, for example. If the CO_2 price is higher than the additional cost of SAF compared to JET A-1, then this emission reduction is also associated with a "negative cost" because the additional cost of SAF is less than the CO_2 price for 1 ton of CO_2.

Usually, the individual measures are ranked by cost and by the amount of CO_2 avoided. In the graphical representation, the zero-cost line represents the marginal cost neutral situation. All columns below the zero line reduce CO_2 and are "negative marginal cost". The width of the column visualises the total CO_2 reduction of a measure. Columns above the zero cost line show measures with positive costs. These measures incur costs for each additional ton of CO_2 avoided. Only if the CO_2 price per ton of CO_2 emissions is higher than the positive marginal cost of emissions avoidance is the result neutral for the airline.

The visual representation reflects that the measures are ranked in order of (negative) marginal cost. Logically, an airline will implement the measures in this order, assuming that keeping the marginal cost of CO_2 abatement as low as possible is an economic principle of cost control (Figure 5.8).

Based on the visualisation, the airline will be able to determine which measures can achieve which amount of CO_2 reduction and what the costs are. A distinction needs to be made between those costs that are one-off and make a permanent contribution to reducing emissions (e.g. reducing consumption by buying a new aircraft) and those measures that have a temporary effect and therefore need to be implemented on a recurring basis in order to achieve a CO_2 reduction in a financial year (e.g. using ground power units at airports instead of operating the aircraft's auxiliary power unit (APU) during the ground time between two flights). From a business point of view, these are costs or expenses that are offset on the credit side by the reduction in expenses for the purchase of CO_2 certificates, especially since – as mentioned above – from 2026 onwards, no free certificates will be allocated for all emissions on intra-European flights, and therefore, the purchase of the required number of certificates will be mandatory, unless the airline purchases CO_2 certificates from the

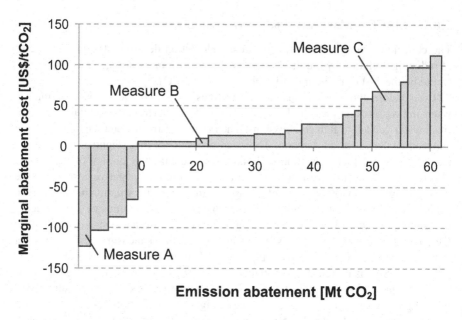

FIGURE 5.8   Visualisation of marginal abatement cost curve (prepared by author).

blending mandate fuel supplier and/or voluntarily purchases additional quantities of certified SAF, which can then also be credited in the EU Emissions Trading Scheme.

It is important to note that the allowance trading will take place in parallel with the regular supply of fossil JET A-1 to European aviation. The break-even point would be reached when the expenditure for the purchase of fossil JET A-1 plus the purchase of CO_2 certificates for this quantity is equal to the purchase of SAF, whose emissions balance will be counted as 100% carbon-neutral fuel in the EU-ETS from an emissions reduction of 70% by legal definition. JET A-1 will tend to become more expensive (even adjusted for inflation) because the number of refineries producing JET A-1 in Europe is declining and this will influence the market price (even if production costs remain unchanged). At the same time, airlines are hoping that SAF will become cheaper to produce on a large industrial scale. As long as the SAF supply in the market is not sufficient to ensure full market supply, SAF prices will remain high as long as all manufacturers can sell their SAF volumes in the market at this price level. As a result, it is likely that SAF will maintain a higher price level than fossil JET A-1 until 2050. The coexistence of different price levels for identical product characteristics in operational use is thus the consequence of the EU Commission's CO_2 pricing. For reasons of profit absorption, it is therefore expected that market prices for SAF will rise to a similar extent as CO_2 pricing in the European Union. From the airlines' point of view, this market reaction, which will affect the blending quotas for intra-European flights, should be taken into account in medium-term financial planning.

With the help of the self-generated emissions balance up to the year 2050, each airline can forecast its probable emissions volume per year. Of course, this requires assumptions about the fleet renewal and expansion that will take place during this period due to market growth. IATA itself concludes that part of the remaining CO_2 emissions can only be offset in other sectors and that the use of SAF will be limited to 65%–70% of the emissions balance. It should be noted that the airline industry relies on "technical innovations" that will reduce fuel consumption per TKO, without specifying which technical innovations are meant and how each development will contribute to reducing consumption. In this respect, it is logical for the airlines to assume that SAF will reduce 2/3 of their CO_2 emissions, assuming a 100% credit, although the net emission reduction value will only be around 70%–80% on average. Therefore, the issues of "maximum allowable blend ratio" and "aromatics content" can be neglected in the economic analysis.

In order to reduce the uncertainty of the assessment, it is useful to present it in three scenarios when assessing the necessary expenditure for constant emission reductions up to the net-zero target in 2050. In terms of market development and domestic fuel demand, business as usual (BAU) should be considered first. This assumes the continuation of the existing fleet with its current consumption levels and ignores market growth. Although this variant is not very likely, it provides the quantitative framework for a continuation of the status quo with rising CO_2 prices and increasing blending rates. For this unrealistic option, the cumulative emissions compensation expenditures can be collected up to the year 2050. For the investment calculation, the present value of the expenditure for 2023 would then be a meaningful figure by discounting the annual expenditure from 2024 to 2050, using the interest rate for long-term loans.

The airline then develops three scenarios from its analyses: the real case scenario reflects the data collected and is considered as an expected reflection of the real development. A worst-case scenario is based on the assumption that expenses will be higher than expected due to less relief from manufacturer delays in aircraft delivery (e.g. Boeing 777 and 787) and higher than currently known CO_2 certificate prices. As a result, achieving the net-zero goal becomes more complex and costly than in the Real Case. However, in the optimistic case, additional aircraft orders are placed for an expanded fleet rollover and CO_2 prices also remain stable. In addition, SAF prices decrease in this model, making the integration of additional SAF volumes beyond the blending mandates economically justifiable for the airline.

With these three scenarios, the airline can assess the range of emission reduction efforts up to 2040 or the target year for net-zero 2050.

In this assessment, mandatory measures should be presented separately from voluntary measures. With these options for action and the range of future abatement costs, nothing stands in the way of defining an emissions reduction path.

5.6.2 SCIENCE-BASED TARGET (INITIATIVE)

As a framework for implementing an airline's decarbonisation strategy, the science-based target (SBT) initiative is one of the possible options, the main features of which are presented below for ease of understanding. The SBT Initiative is a non-profit organisation that has attracted international attention since its inception in 2021. Its founders are the World Wide Fund for Nature (WWF), supported by the International Council for Clean Transportation (ICCT) and the Boston Consulting Group (BCG).

Of course, it is up to each airline to decide how to implement its decarbonisation strategy. However, the SBT initiative offers the advantage that each airline participating in the initiative has the opportunity to benchmark its own emissions reduction efforts against other participating airlines. The methodological approach therefore first requires the definition of achievable targets over time. These targets must be challenging enough for the airline's management to want to and be able to achieve the interim targets through self-defined measures. The emission reduction targets must also be consistent with the budget, and ultimately, the airline's CFO is responsible for financing the measures to be implemented. The targets, their implementation and achievement are methodically tracked and controlled by the airline. In this way, the airline embarks on an emissions reduction path that starts at the board level and must be supported and realised by all levels of management. The scientific, methodical approach requires the will to implement, regular monitoring of success, and an analysis of deviations and their causes if a milestone has not been reached. Various models are known from strategic business planning (portfolio analysis, life cycle curve, etc.), whose complex data are aggregated in a simple and understandable visualisation. By regularly updating these visualisations, changes and one's own position can be communicated in a timely manner. Such TARGET-ACTUAL deviation analyses form the basis for determining the course of action in the next period (Figure 5.9).

To make it easier to track emission reductions, the SBT initiative recommends converting the emission values into a so-called "carbon intensity" as a measure of the emission situation.

The year 2019 can also be used as a baseline, as the years 2020, 2021 and 2022 will not have realistic baseline values for air transport due to the traffic restrictions caused by the pandemic of Covid19. Under the premise of aiming for realistic emission reduction targets, the first inventory should map the expected emissions in the coming years and the measures required to achieve a balanced situation, in order to prevent the airline's individual CO_2 emissions from rising above the historical value for 2019 as a first step. It is possible that the maximum value can already be reached by replacing old aircraft with new ones. In any case, this emissions calculation is

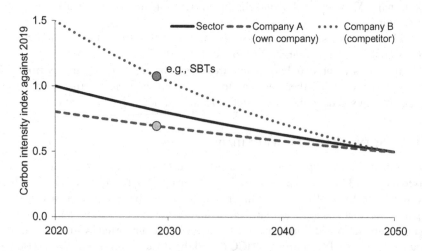

FIGURE 5.9 Illustration of carbon intensity indexed against 2019 (prepared by author).

a useful exercise to prepare for further emissions reduction steps. Of course, the CO_2 offsets required by the EU-ETS can currently be offset as emission reductions through the purchase of allowances. The airline's emissions situation in 2025 would then be the conversion of the "carbon intensity" from 2019 to the new value of the forecast for 2025. This data collection must be done on an annual rolling basis, with SAF purchases – both regulatory and voluntary – included in the emissions reduction.

Instead of costly estimations of the emissions situation of the respective competitors, the airlines can also communicate this data among themselves on a voluntary basis without violating competition law. As soon as the path of planned and progressive emission reductions is followed, the achieved successes are suitable for reporting in the media or in the airline's annual report.

As a consequence of the emission reduction path, medium- and long-term financial planning must include the necessary expenditures for new aircraft, SAF purchases or offsetting through the purchase of certificates. In view of the competitive situation, it therefore makes sense to consider a possible industry-wide reduction curve as a reference point at the beginning of the process, in order to be able to evaluate one's own measures in the context of the market average of all airlines participating in the SBTi.

In the worst case, this approach could result in the airline being above the industry average and producing more CO_2 emissions than the average of the SBTi participants. This "uncomfortable" situation should be a good reason to make additional efforts to maintain environmental competitiveness.

5.6.3 CREATE A PLAN B

Responsible planning always includes an alternative plan, called "Plan B". As far as the emissions situation in the transport sector is concerned, other modes of transport such as cars and railways can now point to declining emission levels. This is mainly due to electric mobility in private transport, as well as the electrification of railway lines and the supply of green electricity from renewable sources to rail transport. In contrast, emissions from civil aviation increased until 2019 and are expected to be at the same level in 2024 as in 2019. The key question for the future of civil aviation is therefore: what is the future of aviation if global CO_2 emissions increase in absolute terms as a result of the balance of emissions reductions from the measures described above and emissions increases from aviation growth? In other words, "What is the airlines' Plan B if they do not reduce their emissions on the scale required?"

As noted above, the US government is promoting the use of SAFs to drive US decarbonisation. In this situation, the simplest and most convenient solution for the airlines would be to promote even more domestic SAF production and, if necessary, scrapping premiums for old US registered aircraft. The question is whether such an extensive subsidisation of air travel can be financed from the national budget in the long run, and whether there aren't other ways of spending government funds that would lead to better CO_2 reductions. In any case, an insufficient reduction of emissions from aviation in the US could also lead to regulatory measures. One such regulatory measure could be a national flight ban on very fuel-intensive aircraft, which would also apply to foreign airlines seeking to fly to the US.

From a European perspective, the question of a "Plan B" is similar to that in the US, because European airlines are even more dependent on the oil companies for their fuel supply than are US airlines.

The prevailing practice in the EU27 of reducing emissions through regulation opens up far more options for "motivating" airlines to reduce emissions. It is not without reason that the European approaches are referred to as "sticks", while the US incentives are referred to as "carrots".

The strictest form of market intervention in the EU would be the allocation of CO_2 emissions to European-based airlines in proportion to the amount of jet fuel they refuel at European airports. Foreign airlines flying into the EU27 would similarly be subject to CO_2 emission quotas on their refuelling volumes for flights from the EU to third countries. The CO_2 quotas available to each airline would be reduced by a flat rate each year. The airlines concerned – both European and visiting – could then decide whether to use more efficient aircraft, buy SAFs, or reduce the size of their aircraft or the number of weekly flights. In an extension of this scenario, CO_2 sequestration would also be allowed as an option, although this is rejected by many NGOs in Europe as insufficient. In this "Plan B", CO_2 offsetting is unlikely to play a significant role because the number of sustainable projects with realistic CO_2 reductions is limited.

The scenarios mentioned here are to be considered "drastic" in terms of their impact on the economic results of the airlines concerned. On the other hand, this "look into the abyss" is intended to illustrate that a lack of success in reducing emissions could also lead to politicians being forced to take tough measures if the climate situation continues to change in the coming years. The EU's decision to extend the EU-ETS to all international flights from third countries to Europe or from Europe to third countries if the ICAO Carbon Offsetting and Reduction Scheme for International Aviation (CORSIA, see following sections) does not make a sufficient contribution, contrary to the previous willingness to compromise also points in this direction.

So far, the airlines that are members of IATA have not come up with a convincing approach that would at least set emission reduction targets at 5-year intervals until 2050. Instead, both the industry association IATA and the airlines themselves remain noncommittal.

Conclusion: For airlines, jet fuel is more than just a cost factor in the operating costs of an aircraft. The availability of jet fuel is the existential foundation of the airline business model: moving passengers and cargo safely from one place to another. But validity of this business model is eroding because climate change is a new variable in this established transportation sector, and its impact on the future of aviation is not yet being taken seriously by all decision-makers. Management, in particular, is expected to make decisions with the future of the company in mind, not just short-term profit optimisation. Shareholder value includes value creation in the best sense of the word. In an era of transition from fossil energy production to climate-friendly energy supply, this is a balancing act for which the aviation industry should be well prepared. A significant part of the global economy depends on air transport. Airline managers should take this into account in their decisions. Airline CEOs should adopt the FORDEC (Facts, Options, Risks, Decision, Execution and Control) structured decision-making methodology developed for pilots to decouple personal decision-making authority from their own preferences (Hoermann, 1994).

6 The Impact of International Regulations on Feedstocks and SAF Production

The fundamental economic approaches of the last century came from the economists John Maynard Keynes (1883–1946) and Milton Friedman (1912–2006). While John Maynard Keynes, as the founder of fiscal policy, saw the state in a balancing role, not only compensating for cyclical fluctuations through government deficit spending but also exercising an economic steering function in order to steer markets, Milton Friedman pursued an exactly opposite model of supply-side policy, in which the state should stay out of market developments and leave market forces free to operate. While Keynes saw government debt as a corrective to business cycles, Friedman advocated controlling the money supply as a means of limiting inflation (Wikipedia: John Maynard Keynes; Milton Friedman, 2023) (Table 6.1).

In aviation, Friedman's theory of limited government led to deregulation by US President Ronald Reagan and UK Prime Minister Margaret Thatcher, allowing any airline to fly on any domestic route without restriction. This model was adopted by the EU and today allows all EU-based airlines to fly without restriction on all intra-European routes. The consequences of deregulation in the 1980s were the creation of many new airlines and the emergence of the so-called low-cost airlines, which began with the creation of People Express in the US in 1981. Deregulation led to a massive expansion of capacity and a corresponding drop in ticket prices. Ruthless competition in the industry led to the bankruptcy of several traditional US airlines such as Pan Am, TWA and Eastern Airlines, while American Airlines, Delta Airlines and United Airlines emerged from the consolidation phase stronger after acquiring competitors. In Europe, the financially strong airlines Air France (partly state-owned), British Airlines (privatised) and Lufthansa (privatised) maintained their positions, while well-established airlines such as Alitalia (Italy), Olympic Airways (Greece), Sabena (Belgium) and Swissair (Switzerland) had to cease operations.

In the aftermath of deregulation and privatisation, environmental policy in air transport fell off the radar screen of decision-makers and was treated as a stepchild. With the rise of green parties in Europe, their entry into parliaments and their influence on environmental legislation, it was simply overlooked that it was only a matter of time before government environmental policy, in the sense of Keynesian interventionism, would also turn to the environmental issue of "aviation emissions". The answer was – in continuation of Friedman's theory – to organise market mechanisms ("market-based measures") in civil aviation that would keep the state out as

DOI: 10.1201/9781003440109-6

TABLE 6.1

Comparison of Keynes' Fiscal Policy and Friedman's Monetarism

	Keynes (Fiscal Policy)	Friedman (Monetarism)
Targets	Short-term elimination of economic disruptions	Long-term elimination of equilibrium disturbing factors
Approach	Increase in consumption strengthens demand	Strengthening supply by improving production conditions
Actions	Increasing government consumption through public spending (especially during recession periods)	Promoting investments in the private sector while reducing inefficient government spending
Result	More government spending and less market	Self-regulation of the markets with withdrawal of government influence

Source: (Summary by author).

a regulator and let the market forces of supply and demand control the monetary transfer payments. In response to the Kyoto Protocol, emissions trading was born, allowing the airlines to develop their business model indefinitely and to initialise a so-called "carbon-neutral growth 2020" through compensation payments, in which the emission levels achieved up to that point would remain "penalty-free" and the emissions resulting from further traffic growth would only be compensated by offsetting in other sectors through the purchase of emission reduction certificates from environmental projects such as, but not limited to, the Clean Development Mechanism (CDM).

The competition between the economic theories of Keynes and Friedman is reflected in Europe specifically in a demand-oriented economic policy (promoted by the Social Democratic, Left and Green parties) versus a supply-oriented economic policy (promoted by the Conservative, Liberal and Republican parties). EU environmental legislation is therefore a steering instrument, which prescribes the steering of environmental policy through laws and sanctions and sets specific limits or thresholds for this purpose. In contrast, US legislation is based on the theory of expanding supply by creating attractive market conditions that allow market forces to develop freely. Of course, the US government also has specific goals that are set out in legislation, but the goal is to achieve environmental goals by controlling market incentives and, if necessary, adjusting them if the expected market response is inadequate. While in Europe, environmental legislation is increasingly interfering with business activity, forcing companies to implement the policy desired through penalties or loss of subsidies, the framework set by the US government allows companies the freedom to choose the method that leads to the desired result.

Air transport has been able to develop well in this field of tension between system theories because its economic systemic relevance has so far provided it with good protection against far-reaching state intervention. In addition, the reference to the consistently meagre annual results, which in turn were a consequence of unrestrained competitive pressure, helped. By arguing that they lacked the financial strength, it was possible to ensure that the EU's ETS measures could be borne by the airlines without economic distortions due to the massive allocation of free emission rights, and that these measures would affect all EU competitors equally, thus creating a balance of power. With the strengthening of Middle Eastern airlines, which massively expanded their activities to capture EU-Asia traffic, the competitive situation changed. The previously high-margin long-haul flights suddenly came under price pressure, as the European airlines (with the exception of Singapore Airlines and Qantas) did not perceive their Asian competitors as a threat to their business model due to quality differences. The Gulf carriers were able to gain significant market share thanks to their modern fleets, excellent in-flight service, numerous extras and, above all, very attractive fares, which, however, required a stopover at the airline's hub. In addition, with their "mega-hub" business model, they were able to offer travellers a wide range of destinations that far exceeded the flight programme of a European legacy carrier. From then on, the European airlines were in for a real shock: European regulation caused additional costs due to ever-increasing environmental regulations and at the same time eroded the price basis in the so important revenue segment of the EU Far East long-haul segment. European airlines demanded that their governments treat all airlines equally and include all international flights in the EU-ETS. As a result, EU legislation stipulated that all flights to airports in the EU should be subject to EU CO_2 compensation through participation in the EU-ETS from the foreign departure airport. The same requirement should apply to flights departing from a European airport to a foreign destination airport.

This unilateral requirement by the EU Commission led to massive protests from the US, China, Brazil and India, which not only threatened to challenge the regulation in court at the United Nations but also threatened to boycott the purchase of European-made aircraft, which would have put the European aircraft manufacturer Airbus in massive turmoil. For the time being, the EU Commission has abandoned its plan to extend the EU-ETS ("stop the clock" agreement) in exchange for an agreement to establish a global emissions trading system under the auspices of the ICAO.

6.1   ICAO AND TAXATION REGULATIONS FOR JET FUEL

In 2010, the 37th ICAO Assembly adopted a resolution requiring all airlines to achieve "carbon-neutral growth" by 2020. The programme is called "CORSIA", which stands for "Carbon Offsetting and Reduction Scheme for International Aviation". Airlines will be required to offset additional kerosene consumption above their 2019 baseline kerosene consumption, which should remain uncompensated. The compensation must be made by offsetting CO_2 emissions through the purchase of emission reduction certificates, or it may be made through the purchase of SAF volumes that meet the sustainability requirements of ICAO as the common regulation for CORSIA eligible jet fuel. Again, in the case of CORSIA, the global jet fuel volumes burned on international "cross border" flights remain untouched for the time being and will not require CO_2 compensation through offsetting or SAF purchase until 2027!

ICAO decisions are not directly enforceable by the airlines concerned. Rather, it is up to each individual Member State to adopt the ICAO decisions into national law and thus order their implementation within its national territory. There is no fixed timeframe for the implementation of the resolutions. Each government is responsible for the timely implementation of the resolutions. The specification of a fixed implementation date in an ICAO resolution therefore does not mean that the decision will be implemented into national law by all governments worldwide by that date.

As noted above, when the ICAO was established, it was agreed internationally through a resolution that no fuel tax or VAT would be levied on international flights. This resolution does not apply to domestic flights, although many states exempt domestic flights from taxation for practical reasons. In some countries, national governments have now developed other sources of taxation that tax both domestic and international aviation without violating the ICAO non-taxation rule. In some European countries, for example, an "environmental aviation tax" is levied, which is graduated according to the distance flown and charges airlines a fixed fee per passenger, which they include in the calculation of the ticket price like any other expense. A total of six European countries levy an aviation tax. The tax applies to the entire flight from the origin to the destination airport, so the distance class does not end at stopover or transfer airports. Passengers who only transfer at an airport within the six participating EU countries do not pay the tax. Feeder flights are also not taxed separately; the tax rate for the destination applies.

The majority of flights departing from Germany on which air traffic tax is paid fall into the lowest distance class (83% in 2019). 5% of passengers paid the medium rate and 12% the long-haul rate (Destatis, 2020).

In times of high kerosene prices, airlines have added an additional kerosene levy to the ticket price as a special item in order to be able to reflect the non-calculated kerosene prices in the route profitability calculation in a cost-covering manner. In Europe, European consumer protection legislation requires airlines to display all ancillary fees together with the ticket price in a single amount to provide passengers with greater price transparency and enable them to choose the airline of their choice based on price level.

When looking at fiscal activities in Europe, it should be noted in a value-neutral manner that all aviation taxes, including those with an alleged environmental link, flow as tax revenue into the respective national budgets. There is no ecological earmarking of the revenues. In this respect, there is no positive correlation between the additional revenues in the state budget and the state expenditures for climate and environmental protection. Rather, it is argued that the increase in ticket prices alone would reduce the growth of air traffic, since the household income of low-income passengers would not allow for further air travel, and thus, fewer citizens would book and travel by air as a result of the tax levy. Apart from the fact that such a policy would again reserve air travel for wealthy passengers, booking behaviour after the end of the Covid19 flight restrictions shows that flight bookings remain price inelastic even in recessionary periods: Air travellers save elsewhere in their household budgets, but do not forgo private vacation travel. This finding raises the question of whether environmental protection is really the guiding rationale for additional levies, or whether the introduction of indirect taxes is not primarily intended to provide additional revenue to national budgets. A comparison of the amount of environment-related levies on the revenue side of a national budget and an explanation of what amount is clearly spent on climate and environmental protection on the expenditure side would at least provide more transparency on this issue

6.2 EU-REGULATION

6.2.1 EU-ETS

Beginning in 2012, European airlines were integrated into the European Emissions Trading Scheme (EU-ETS), which requires them to purchase CO_2 emission allowances for every ton of CO_2 emitted on flights within Europe. As an incentive to airlines, the EU Commission issued free allowances for each of the trading periods, which range from 5 to 8 years. The allocation of free allowances is based on each airline's individual mileage data, but at a glance, EU airlines received 85% free allowances in 2012, 82% free allowances in the third trading period 2013–2020, and from 2021, 82% in 2021, decreasing to 75% in 2024 and 25% in 2025. Free allowances will end on 31 December 2025. Airlines that need more allowances than they receive for free will have to participate in auctions and can buy allowances.

In effect, the ICAO Carbon Neutral Growth Regulation and the need to purchase additional allowances under the EU-ETS can be aligned as ICAO regulations transposed into national law.

It should be noted that the high-fuel-consuming European legacy carriers use about 75% of their jet fuel on long-haul international flights and only about 25% of their jet fuel on intra-European routes. The vast majority of their annual consumption is therefore not subject to the EU-ETS! This is not the case for European regional airlines and European low-cost carriers, which operate mainly within the EU-27, with almost 90%–100% of their flights within Europe, but their share of jet fuel in aircraft operating costs is significantly lower than that of the legacy carriers.

6.2.2 EU-RED II

The Renewable Energy Directive (RED) was originally introduced by the EU Commission in 2009 (Directive 2009/28/EC, dated 23 April 2009) and set a target of meeting at least 20% of gross final energy demand from renewable sources by 2020. This improvement in emissions was to be achieved by replacing fossil fuels with renewable energy sources. As a result, the EU-27 Member States achieved a 22% reduction in their emissions in 2020, meeting the target set in 2009.

In 2018 – before the end of the period set in the original directive – the EU-RED was amended (Directive 2018/2001 EU) and adopted on 11 December 2018. According to this directive, at least 32% of the gross final energy consumption in the member states must be replaced by renewable energy by 2030. An essential part of the amendment was the promotion of electricity generation from renewable sources (solar, wind, hydro). For the transport sector, it was stipulated that at least 14% of energy consumption must come from renewable energy sources by 2030, which includes in particular the fuel sector. Thus, the minimum requirements (threshold) for renewable fuels to be counted in the EU-ETS have been raised to 70% greenhouse gas savings compared to the fossil substitute as of 1 January 2021. This value also applies to renewable fuels of non-biological origin (RFNBO) for transport. The background to this is the indirect consideration of an "indirect land use change" (ILUC) factor, which is intended to compensate for the negative effects of expanded land use for energy crops on food production. Although the displacement effect on food production areas assumed for ILUC could not be empirically proven despite various analyses and research approaches, the EU Commission has taken into account possible land competition by increasing the minimum requirements.

In addition, the GHG savings must also be achieved through the use of "advanced biofuels" in the amount of 1% in 2025 and 3.5% in 2030. According to Annex IX, Part A, these are wastes and residues such as straw, wood residues, bagasse, manure and municipal solid waste (MSW), but excluding household waste used for the recovery of recyclable materials.

6.2.3   EU-RED III

With the agreement of the European Council (consisting of the heads of government of the EU27 member states), the European Parliament and the European Commission (comparable to the cabinet of a sovereign state with ministers and the head of government, appointed commissioners and the president of the EU Commission) in the so-called "trialogue negotiations", the existing EU-RED II Directive will be amended once again.

Instead of the previous 32.5% GHG reduction, 45% of energy demand must now come from renewable sources by 2030. The negotiations took longer than expected because the compromise already reached had to be renegotiated, as France demanded that nuclear power be recognised as "green" energy under the future EU-RED III Directive. With the acceptance of this demand, the "trilogue" negotiations ended on 16 June 2023. For the final conclusion of the negotiations, the European Parliament still has to approve the wording of the Directive.

The future RED III Directive (which has not yet been published in the Official Journal of the European Union for the reasons mentioned above) will for the first time include air transport in the mandatory emission reductions in Europe. For the transport sector, the previous GHG reduction target of 14% by 2030 has been increased to 29% by 2030. In the aviation sector, EU member states must require distributors of JET A-1 to blend a minimum amount of SAF into JET A-1 starting in 2025: 2% in 2025, 6% in 2030, 20% in 2035, 32% in 2040, 38% in 2045 and 63% in 2050. In addition, a sub-quota of 1.2% for electricity-based jet fuel from hydrogen (eSAF) must be included in the JET A-1 from 2030. This increases to 2% in 2035. The eSAF volumes count towards the minimum SAF quota to be achieved.

In Germany, a mandatory eSAF sub-quota of 0.5% from 2025, 1% from 2028 and 2% from 2030 has already been in force since 2021, which was legally anchored by the German government on the basis of the PtL roadmap (BMDV, 2021).

6.3   EU POLICIES – MANDATES AND PRODUCTION INCENTIVES

6.3.1   REFuel EU

In March 2020, the EU Commission launched the REFuelEU initiative, a programme to promote advanced biofuels and electricity-based fuels. As part of the European Green Deal, targets will be agreed to reduce emissions from transport by 90% by 2050 (compared to 1990) and to increase the production and use of sustainable alternative fuels for transport. The REFuelEU initiative aims to increase the production and use of sustainable alternative fuels for different transport modes, including aviation.

The EU Commission has shortlisted several activities as areas for policy action:

1. SAF blending mandate: This objective is already being implemented with the amendment of EU-RED III.
2. Revision of the multiplier: In the context of crediting SAF towards the sectoral target, a multiplier of 1.2 instead of 1.0 could incentivise Member State support measures.
3. A centralised auctioning mechanism – as already practiced in the auctioning of renewable electricity – could allow SAF suppliers to sell fixed quantities of SAF through auctioning procedures.
4. Financing mechanism – the EU could channel and broker appropriate funds under EU auspices to fill the financing gaps for SAF production sites.
5. Prioritisation – allowing rare commodities to be reserved specifically for SAF production.
6. Voluntary agreements – where appropriate, the EU would provide a platform for SAF suppliers and buyers to negotiate contracts and improve market transparency.
7. Technical facilitation and support initiative – through this offer, the EU would provide assistance to new SAF producers in the formal project application and implementation process.
8. Monitoring to document and control SAF production and demand.

REFuelEU also addresses the issue of carbon leakage through tankering, i.e. taking additional fuel from a non-EU airport in order to minimise the amount of fuel to be refuelled at the EU airport, as the additional cost of emissions taxation will increase the fuel bill of the responsible airline.

According to REFuelEU, the amount of fuel to be refuelled on a flight should include at least 90% of the amount that would be refuelled under normal operating conditions for the upcoming flight, assuming that the inbound flight has fuel reserves on board for holding patterns, missed approaches and onward flights to an alternate airport with normal operating fuel reserves.

However, it is currently not clear how the EU Commission intends to implement its own proposal, as the pre-calculated fuel quantity for a flight can be increased by the captain, as the ultimate decision-maker, on his own responsibility in view of the weather situation or possible airspace closures, in order to ensure the safe execution of the flight even under exceptional conditions.

The EU Commission's DG Move is responsible for implementing the REFuelEU initiative. The initiative has the goal of an independent regulation, but is initially seen as a support for the tasks resulting from the various directives. REFuelEU will not become a directive that has to be transposed into the national law of the EU member states (EU-DG Move, 2020).

6.3.2 Fit for 55

Under the working title "Fit for 55", the EU Commission intends to combine all legislative projects up to 2030 in a single package. Fit for 55 means a 55% reduction in emissions compared to 1990 levels. The Fit for 55 package is part of the EU Green Deal, which aims to make the European Union carbon neutral by 2050. In this context, the EU Commission has bundled several measures that will be implemented in the form of legal requirements. These include (1) expanded emissions trading, (2) a phase-out of internal combustion engine cars and (3) a CO_2 border adjustment to prevent companies from relocating abroad to avoid EU emissions taxes. The RED III agreement already implements parts of the Fit for 55 programme, such as the elimination of free allowances for aviation (Figure 6.1).

6.3.3 Price Imparity Jet A-1 versus HEFA – SAF in the EU

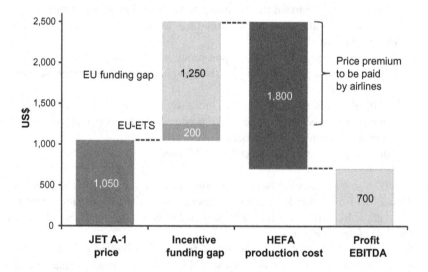

FIGURE 6.1 EU Price imparity JET A-1 and HEFA in USD/t (acc. to prevailing EU-ETS scheme in 2023) (prepared by author).

6.4   US REGULATIONS

6.4.1   ROADMAP FOR SUSTAINABLE AVIATION FUELS (GRAND CHALLENGE)

The US Departments of Energy, Transportation, Agriculture and Defence, as well as NASA, the General Services Administration and the Environmental Protection Agency (EPA) are working together on this US government initiative launched on 9 September 2021 called the SAF Grand Challenge. The goal is to reduce today's business-as-usual aviation emissions by 20% by 2030 and achieve a carbon-neutral aviation sector by 2050 by increasing SAF production capacity to 3 billion US gallons per year by 2030, based on a total of USD 297 billion to be made available over a 5-year period ending in 2030. The target year for 2050 is 30 billion US gallons of SAF per year. USD 4.3 billion is available for SAF production facilities under construction and planned. A research programme aims to increase aircraft fuel efficiency by 30%. In addition, the airspace structure and airport operating facilities will be reviewed for efficiency improvements: The goal is to improve air quality and reduce emissions at and within airports. At the same time, the programme will support rural areas through increased demand for biomass feedstock for SAF production, creating new and well-paying jobs there and in the manufacturing industry. At the same time, support for SAF production will set the minimum threshold for GHG reductions at 50%.

EPA and DoE will also increase their collaboration to evaluate new SAF production processes and SAF feedstocks for participation as renewable fuel producers with Renewable Identification Numbers (RINs) in the federal Renewable Fuel Standard programme.

6.4.2   INFLATION REDUCTION ACT OF 22 AUGUST 2022

With the Inflation Reduction Act (IRA), the US government enacted two additional legislative initiatives to promote SAF production at the federal level:

The Sustainable Aviation Fuels Credit Programme (commonly referred to as the Blender's Tax Credit (BTC)) will be paid to blenders in 2023 and 2024 at a rate of USD 1.25/USG SAF if the SAF produced meets the minimum requirement of a 50% GHG reduction. If the LCA GHG quota is above the 50% minimum, an additional USD 0.01/USG will be paid for each additional percentage of emission reduction, up to a maximum of USD 1.75/USG for 100% emission neutrality. The Sustainable Aviation Fuels Credit Programme is limited in time to the years 2023 and 2024 (Sections 40b and 6426(k) of the Internal Revenue Code).

For the years 2025 through 2027, the technology agnostic Clean Fuel Production Credit applies pursuant to §45(z) of the Internal Revenue Code. The amount of the subsidy is based on the amount of GHG reduction as determined by the GREET life cycle analysis. The lower the carbon intensity score, the higher the subsidy. The maximum possible amount of 1.75 USD/USG corresponds to the maximum amount of the Blender's Tax Credit, but the assessment is backward-looking, using the emission factor of 50 kg/CO_2 eq per MMBtu as the maximum value for participation in the programme.

Both credit options can be used to offset excise tax liability or as a direct payment in the event of insufficient excise tax liability. The tax credit is subject to additional eligibility requirements.

Overall, the IRA-related SAF tax credits for SAF will replace the current Blender's Tax Credit, which is capped at USD 1.00/USG and can be stacked with Renewable Fuel Standard 40 RIN credits and Low Carbon Fuel Standard (LCFS) credits applicable in the states of California, Oregon and Washington. Specific terms and conditions of the current BTC apply.

6.4.3   State Level SAF Incentives

6.4.3.1   California Low Carbon Fuel Standard (LCFS)

The State of California enacted the LCFS in 2009 to reduce greenhouse gas emissions from the transportation sector. The goal is to reduce the carbon intensity of transportation fuels used in California and improve the state's air quality. It also aims to reduce the state's long-term dependence on fossil fuels. The regulation was amended in 2019 to recognise SAF as an LCFS-eligible fuel for state tax credits.

The GHG reduction potential of SAF will be determined and quantified through a life cycle analysis of SAF. The benchmark for the carbon intensity of SAF is the amount of emissions reduction compared to commercial JET A-1, as measured by the Carbon Intensity Metric (California Air Resources Board, 2023) (Figure 6.2).

6.4.3.2   Oregon Clean Fuels Programme

The State of Oregon implemented the Clean Fuels Programme in 2016, which is administered by the Department of Environmental Quality Commission (DEQ). Based on the 2016 baseline, gasoline, diesel, and, on a voluntary basis, jet fuel must reduce their carbon intensity by 20% by 2030 and by 37% by 2035. DEQ continually reduces the allowable carbon intensity of fuels each year to meet the annual reduction targets. Fuels that are less carbon-intensive than the annual limit generate credits, while fuels that are

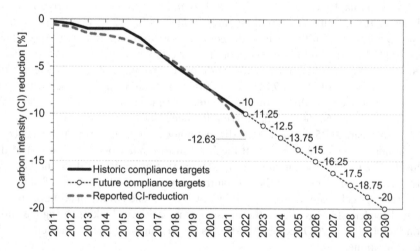

FIGURE 6.2 Performance of LCFS 2011–2021 according to California Air Resources Board (2023) (prepared by author).

more carbon-intensive generate deficits. Credits and deficits are measured in metric tons of greenhouse gas emissions. Participants in the programme can sell credits to offset deficits, which in turn generates revenue to pay for projects that reduce greenhouse gas emissions. A surplus of credits can be saved to offset future deficits the entity may face or for future sale, assuming demand for such credits increases (DEQ, 2023).

6.4.3.3 Washington State Clean Fuels Programme

The Washington State Department of Ecology has successfully implemented the Clean Fuel Programme. State law requires fuel suppliers to gradually reduce the carbon intensity of transportation fuels to 20% below 2017 levels by 2034.

The programme is open to voluntary participation by SAF suppliers, but participation is currently focused on gasoline and diesel fuel suppliers (Department of Ecology, 2023).

6.4.3.4 Illinois Sustainable Fuels Tax Credit

State lawmakers in Illinois have passed legislation to create a $1.50/USG tax credit for sustainable aviation fuel. Domestic airlines may use the tax credit to satisfy all or a portion of their state use tax liability. The legislation was signed into law on 3 February by the Governor of Illinois and has become law.

The tax credit will be paid for each gallon of SAF that meets the GHG reduction threshold of 50% less emissions than conventional jet fuel sold to or used by domestic airlines that refuel their aircraft in Illinois for a US domestic flight between 1 June 2023 and 1 June 2033. Airlines fuelling aircraft for international flights are not eligible to participate in the programme.

Airlines operating to and from Illinois may claim the SAF tax credit up to a certain amount based on the amount they spend on domestic jet fuel in that year. Airlines may also combine the Illinois credit with the new $1.25–$1.75/USG credit under the federal Inflation Reduction Act (Biodiesel Magazine, 2023).

6.4.4 CARBON CAPTURE, UTILISATION AND SEQUESTRATION (CCUS)

As a supporting measure to meet the minimum GHG mitigation requirements, CO_2 process emissions from SAF production can be captured and stored underground. This improves the GHG mitigation value to the extent of the CO_2 storage, although the emissions from the storage process in turn worsen the GHG mitigation value.

As defined by the EU, carbon capture and geological storage (CCS) is a bridging technology that contributes to climate change mitigation. Carbon dioxide (CO_2) is captured from industrial installations and transported to a storage site, where it is permanently stored in a geological formation suitable for permanent storage (EU 2009/31/EC). The CO_2 to be stored can come from fossil-fuel power plants, from industrial plants or from the use of biomass for energy purposes. Storage is permitted in depleted gas or oil reservoirs, saline aquifers or marine subsoil, provided that the nature of the subsoil formation ensures safe storage without diffuse CO_2 leakage into groundwater bodies or to the surface.

Although research and pilot projects have confirmed that CO_2 can be safely stored, industry has long been reluctant to embrace this option because the cost of storage is higher than the amount of carbon credits it can generate. In this respect, carbon offsetting has an undesirable side effect that is often ignored when considering market-based measures: As simple and convenient as a financial transaction mechanism is, it also prevents direct measures that reduce CO_2 in the atmosphere if the marginal cost of CO_2 avoidance is higher than the marginal cost of financial offsetting.

Fortunately, this situation has changed in the early 2020s: The US federal government announced significant CCUS opportunities, including new funding in the Infrastructure Investment and Jobs Act of 2021 and favourable CCUS tax credit changes in the Inflation Reduction Act of 2022. The EU proposed a Net Zero Industry Act on 16 March. On 16 March 2023, as part of the European Green Deal, the EU proposed an annual CO_2 injection target of 50 Mt CO_2/year for 2030 and improved permitting procedures for CCUS. In addition, the pilot phase of the Greensand project in Denmark became operational in March 2023, transporting CO_2 from Belgium for storage in a depleted oil field in the Danish North Sea (IEA, 2023).

While CCS is recognised and practiced in the US as an option to improve GHG emissions, the EU has long been passive in this regard. In Germany, a successfully operated pilot plant near Berlin (Ketzin, Brandenburg, 2004–2017) confirmed the safe and sustainable storage of CO_2. The use of a potential storage formation is based on the following prerequisites: (1) sufficient storage capacity (absorption capacity), (2) the presence of a geological barrier (caprock) that safely seals the reservoir over a long period of time and remains stable even at elevated pressures, and (3) predictable interactions of the CO_2 with the reservoir that are conducive to storage. Under these conditions, CCS becomes an important part of aviation's net-zero strategy (Figure 6.3).

6.4.5 Price Parity Jet A-1 versus HEFA-SAF in the US

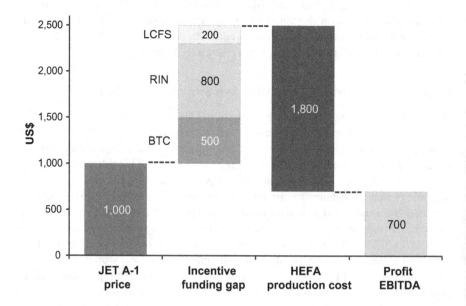

FIGURE 6.3 US Price Parity JET A-1 and HEFA (acc. to prevailing US government incentives in 2023) (prepared by author).

6.5 CARBON OFFSETTING AND REDUCTION SCHEME FOR INTERNATIONAL AVIATION (CORSIA)

6.5.1 History

In October 2016, the ICAO Assembly decided to introduce a global carbon offsetting scheme for international aviation. This followed the EU's announcement to extend the EU-wide Emissions Trading Scheme to international flights to and from Europe, which led to significant international protests. As a result, ICAO member states agreed to establish their own scheme that would apply only to international aviation. Under this arrangement, domestic flights are not covered by the CORSIA system and remain exempt from compensation.

In order to avoid undue burdens, States whose air traffic volume is less than 0.5% of the world's air traffic will also be exempted from CORSIA. In addition, flights by small aircraft, rescue flights and island supply flights will remain exempt.

With CORSIA, emissions from international flights will be subject to a CO_2 levy in the future (from 2027), but the fuel volume of international airlines achieved by 2019 will also remain exempt from the levy and only the fuel volume above the 2019 reference volume will be used to calculate the levy. As a result, the CO_2 offset model will only build up over time and, taking into account the increase in jet fuel consumption, will cover at best 25% of global jet fuel demand in 2050. This is because of the tax exemption for domestic flights, which account for the majority of transportation, especially in the US, China and India. Assuming a tax-free jet fuel volume of 290–300 million tons of JET A-1 in 2019 and a dynamic annual growth pattern of 2% additional demand compared to the previous year, the expected demand volume in 2050 is approximately 510–520 million tons of jet fuel. Since the reference quantity remains tax-free, CORSIA then refers to the growth of international flights, whose share of the additional demand will be at most 60%. CORSIA thus covers approximately 130 million tons of jet fuel for long-haul flights out of 510 million tons. This corresponds to a share of about 25% in 2050.

The simplest way to measure emissions from international flights would have been to relate them to the amount of fuel used, since airport fuel depots are operated under Customs supervision due to tax exemption, and Customs periodically checks jet fuel consumption. Flight numbers would have provided a quick, reliable and unambiguous measure of fuel consumption and associated emissions.

However, the ICAO member states could not agree on such a simple calculation model because the representatives of the developing countries pointed out that their national airlines predominantly operated older aircraft with corresponding fuel consumption. Conversely, airlines that had invested heavily in new, fuel-efficient aircraft would have been favoured. As a result, a "route-based approach" was agreed, whereby each international route is based on a standardised distance, regardless of the actual routing of the individual flights on that route. The standardised distance is then used to calculate a standardised average fuel consumption for each aircraft type for the respective flight distance. Since such a calculation model still results in higher consumption values for older aircraft, it was agreed that the amount of CO_2 to be compensated would be determined either from the

average consumption of all registered aircraft (sectoral approach) and/or from the
calculated individual consumption of an airline depending on the composition of
its fleet (individual approach). In practice, this encourages airlines to fly old and
fuel-intensive aircraft until the end of their lives, since they only have to bear a
limited share of the immense fuel consumption of older aircraft, at least as far as
CO_2 compensation is concerned.

If two airlines fly an international route and one of them does not participate in
the CORSIA system because of the 0.5% rule, the other airline does not have to pay
the levy on this route either.

The implementation of CORSIA follows a timetable:

1. 2019: <u>Monitoring phase</u> begins
   Beginning in early 2019, airlines from all ICAO Member States will be
   required to monitor and report their CO_2 emissions to ICAO under CORSIA.
2. 2021: Voluntary <u>pilot</u> phase begins
   To date, 88 states are participating in the voluntary pilot phase. They
   cover about 77% of international air traffic.
3. 2024: The voluntary <u>1st phase</u> of CORSIA begins.
   Participation in phase 1 is also voluntary. The growth-related emissions
   of international aviation are offset.
4. 2027: The <u>mandatory phase</u> of CORSIA begins.
   States with a share in international aviation of more than 0.5% in 2018
   are required to participate. Overall, however, at least 90% of international
   air traffic must be covered by CORSIA. If the quota is not reached, the
   countries that have not yet committed themselves must also participate in
   CORSIA.
5. 2032: Review of the effectiveness of CORSIA
   Beginning in 2022, CORSIA will be reviewed every 3 years to assess its
   contribution to the sustainable development of international aviation and its
   effectiveness. In 2032, a decision will be taken on whether CORSIA should
   be allowed to expire in 2035 or whether it should be continued. (This is the
   point at which the EU Commission will intervene to extend the EU-ETS in
   place of CORSIA if the results achieved are unsatisfactory).

Regarding the methodological understanding of air traffic growth, it should be noted
that the agreed rules require that only traffic growth above the 2019 CO_2 emission
level be offset, or, according to the latest adjustment, from 2024, growth above 85%
of the 2019 CO_2 emission level. For this purpose, ICAO annually determines the
"Sectoral Growth Factor" (SGF) as the basis for CO_2 offsets. However, as a result of
the Covid19 pandemic, the CO_2 emission value in 2021 collapses to 167 million tons
of CO_2 compared to the 2019 baseline of 341 million tons. As the achieved CO_2 output from international flights is therefore below the baseline, all CO_2 offset payments
for 2021 are cancelled. This fact is interesting in that it shows the share of international traffic compared to domestic traffic. The 341 million tons of CO_2 of the base

year correspond to 108.3 million tons of JET A-1, whereby intra-European flights are not counted under CORSIA. Assuming an annual consumption of 290 million tons of JET A-1 in 2019, international flights account for about 37%–38% of fuel consumption.

The 41st Session of the ICAO Assembly adopted Resolution A41-22, which strengthened the collective strength for the implementation of CORSIA. After conducting the first CORSIA Periodic Review, the Assembly revised the CORSIA baseline:

"For 2021–2023, 2019 CO_2 emissions shall apply.

For 2024–2035, the baseline is reduced to 85% of 2019 CO_2 emissions.

The percentage of sectoral and individual operator growth factors for the calculation of offset requirements under CORSIA was also revised:

For 2021–2032, 100% sectoral only

For 2033–2035 max. 85% sectoral + min. 15% individual

New entrants shall use 2019 CO_2 emissions to determine CO_2 offsets" (CORSIA Newsletter October 2022).

6.5.2 MARKET-BASED MEASURES

CORSIA is based on the so-called market-based measures, i.e. the purchase of emission allowances by the private sector, which are used to finance emission reduction projects in other sectors, which in turn should lead to corresponding long-term emission reductions.

The design of this trading system is currently under discussion within ICAO. However, it has taken some effort to anchor SAF as an alternative form of offsetting in ICAO. As with the EU-ETS, airlines will be free to choose their preferred method of compensation.

The question of whether there will be enough attractive offset projects worldwide to be financed by airlines through the purchase of certificates cannot be answered unequivocally at present. NGOs fear that, in the absence of a monitoring institution, the funds will be used for projects whose effective CO_2 reduction could turn out to be significantly lower than promised, especially since the airline purchasing the certificates has no obligation to monitor how effectively the funds are used and whether the promised emission reductions are actually achieved (International Transport, EDF, ICSA, 2020).

As an alternative to purchasing certificates for non-sectoral projects, it is possible to purchase CORSIA-eligible aviation fuels. The use of SAF as CORSIA eligible fuel requires separate certification of the SAF and a verified certificate of origin. The international certification organisation "International Sustainability and Carbon Credits" (ISCC), for example, offers a CORSIA-compliant ISCC-CORSIA certification in addition to the EU-ETS-compliant SAF certification (ISCC-EU), which takes into account the specific requirements of CORSIA. ISCC does not carry out certification itself; this task is performed by accredited international auditing companies such as SGS, Bureau Veritas, TÜV and other globally operating companies.

6.6 OTHER NATIONAL INCENTIVES AND REGULATIONS (NORWAY, SWEDEN)

6.6.1 NORWAY

In 2016, the Norwegian state-owned airport company Avinor carried out the first commercial use of SAF at Oslo Airport with 4,000 tons of HEFA-SAF as a blending component as part of the EU-funded ITAKA project. ITAKA (Initiative Towards sustAinable Kerosene for Aviation) was designed to map the entire value chain from biomass cultivation (Camelina sativa in Spain) to aircraft refuelling in Oslo. With the support of the Norwegian government, the airlines SAS, Lufthansa and KLM were selected as contract partners. After evaluating the results, the Norwegian government introduced a mandatory SAF blending quota of 0.5% as of 1 January 2019, making it the first government in Europe to establish a permanent use of SAF (CORDIS, 2019). According to press reports, the Norwegian government is currently considering increasing the quota to 2%, which would bring it in line with the EU-RED III Directive. During the ITAKA project, the SAF blend was physically distributed to all aircraft, with only the CO_2 reduction certificates going to the three airlines mentioned. During the test run at Oslo Gardermoen Airport, questions arose for the first time that have not yet been resolved internationally: (1) How can CO_2 reduction certificates be transferred electronically? In Oslo, the fuel receipts were physically collected and sent to the relevant Emissions Trading Offices in the three participating countries, where the fuel receipts were manually processed to collect the proportional SAF quantity for each individual refuelling. (2) How can SAF produced in third countries outside the EU be accounted for in the EU-ETS? (3) If SAF production in third countries has been certified under EU-RED II, how can double counting of these quantities be safely excluded?

The introduction of the 0.5% SAF quota was easily implemented in Norway under the conditions prevailing at the time, as separately produced batches from NESTE, which did not have its own representation at Oslo Gardermoen Airport, were delivered and invoiced by AirBP as a consortium partner. If Norway, as a non-EU27 country, continues to develop the blending quota in analogy to EU-RED III (subject to final approval by the EU Parliament), it is obvious that the open questions of international accounting and inventory management of emission reduction certificates to the extent of SAF production should be regulated in a uniform and comprehensive manner.

6.6.2 SWEDEN

The Swedish government has introduced a national SAF quota of 1% from 2021, with the aim of steadily increasing to 30% in 2030. Swedish Biofuels, a Swedish-registered SAF research and development company, will build three SAF production plants using Alcohol-to-Jet (ATJ) technology with support from the EU Commission, producing 400,000 tons of SAF per year. SAS – the Scandinavian airline of Denmark, Norway and Sweden – is also planning an eSAF project with Swedish energy company Vattenfall, Shell Aviation and process manufacturer LanzaTech to produce 50,000 tons of PtL jet fuel per year.

In reality, not every project announcement will result in an operational SAF production facility, but it is clear that the political framework for a blending quota provides a good basis for subsequent project development. Whereas in the past the lack of a sales market was the main obstacle to investment, blending quotas provide a guaranteed purchase volume that is absolutely essential for the market in question. Combined with appropriate subsidies from the national government, financial investors can then be found, or large companies along the supply chain for chemical plants and energy companies can participate in these investments.

Conclusion: Considering that SAF is a specified product that is offered worldwide as a commodity of identical quality and presumably at similar market prices and conditions (with the exception of PtL jet fuel, which has its own – fixed – price level due to production costs), it is surprising that there are so many regulations, subsidy requirements and specifications that, taken together, tend to achieve the opposite of what is politically desired in Europe and the US: the decarbonisation of aviation. Given the urgency of limiting global warming, the regulatory framework must be simplified, international cooperation between the EU Commission and the US administration must be intensified, and achievable targets must be defined and harmonised. What is missing is a critical examination of whether all the special paths, individual legislation and funding guidelines will ultimately promote or hinder the market penetration of SAF during the transitional period of the coexistence of JET A-1 from petroleum and JET A-1 from SAF. Subsidiarity implies diversity in order to respond to regional differences with targeted measures. From the government's point of view, it is necessary to examine whether or not there is in fact a disparity in the production and use of SAFs that causes the level of complexity faced by project developers.

Investing in a project requires a special kind of trust: trust in the functionality of the technology, in the ability of the market to demand the product, in the long-term viability of the sales channels and in the performance of the management team. If the policy framework changes at short intervals and there is insufficient investment protection for existing investments, the attractiveness of such projects decreases. Legal regulations should ensure equal opportunities for all market participants and support a politically desired goal. The cautious reaction of the capital market to the far-reaching investments should be reason enough to look for the causes of the slow market take-off and to remedy the weaknesses identified.

7 Feedstocks as Starting Point of SAF Production

7.1 BIOMASS-BASED FEEDSTOCKS

7.1.1 CROPS AND WASTE

The suitability of the biogenic feedstocks results from the reverse view of the process chain from aircraft propulsion to jet fuel processing technology and from there to the properties of the biomass. Based on the need for identical product properties of SAF as a complementary and substitute product for JET A-1, only those raw material processing chains are considered that produce a specified end product (in the context of co-production). The raw material base ranges from plant products (lignocellulosic, oil, starch and sugar containing plants, harvest and processing residues and plant residues), animal residues (slaughterhouse waste, animal fats, animal excrements) and organic excrements to organic waste ("green" municipal waste) and algae. In contrast to road transport, each feedstock must have its own biogenic origin in the conversion process to be used as synthetic jet fuel as certified by ASTM D1655 and ASTM D7566.

Therefore, solid or liquid biomasses that are suitable as SAF feedstocks are those that (1) can be grown and harvested in large quantities (cultivated biomass) or processed as residuals (waste biomass), (2) do not impose special requirements on the growing areas (e.g. soil conditions, irrigation), (3) have a high productivity with the lowest possible land use, (4) have the highest possible energy content (carbon content), (5) are easy to transport and store, (6) are regularly available in a constant quality, (7) are easy to process in the respective production processes, (8) whose cultivation, preparation and processing make economic sense, i.e. at reasonable costs that can be passed on to the consumer, (9) meet the SAF requirements of ASTM D7566, and (10) whose cultivation or processing is accepted or desired for ethical or socio-political reasons (Kaltschmitt et al., 2016).

With the beginning of the use of biofuels in the automotive sector through the legally prescribed blending of ethanol (grain as feedstock) and biodiesel (rapeseed oil as feedstock), domestic agriculture in the countries of the European Union as well as in the US, Canada and Brazil began to cultivate the biomass required for this through intensification of production, renewed use of set-aside land (especially in the EU) as well as through structural changes in the distribution of agricultural production. The resulting utilisation chains for biomass used for energy purposes were established on the basis of secure demand and tax incentives. In this respect, the available energetic biomass potential in Central Europe is almost fully exploited. This results in two problems for the aviation industry and its future biomass demand: (1) If the existing biomass cultivation

DOI: 10.1201/9781003440109-7

potential is fully utilised, biomass for aviation can only be made available for aviation in the respective domestic markets of the airlines through price competition (outbidding competition) in a new direction of utilisation. In view of the lack of taxation of JET A-1, the possibility of government support through tax incentives is excluded. (2) Taking into account the European market price level for bio-resources and the more expensive processing compared to gasoline and diesel, the production costs for SAF lead to prices that are significantly higher than the price level for JET A-1. In the absence of a mandate for SAF, the sourcing of feedstocks at prevailing market prices in the EU is in fact a criterion for excluding SAF from the production of synthetic fuels. The lack of available agricultural land in Europe and the current price level in the EU for cultivated biomass in the EU must ultimately lead to a shift in production to third countries, where – assuming the same sustainability conditions – biomass can be cultivated, taking advantage of economies of scale and more favourable production costs compared to the EU.

The conditions for Central European waste biomass are similar to those for cultivated biomass. Here, too, the only way to meet the demand for aviation residues and waste is through price competition. The raw material base is different for the US, Canada and Brazil: Due to the high export surpluses for grain and vegetable oils in the US and Canada, production volumes can be diverted into different processing chains instead of being exported. Brazil, for example, has a large potential of undeveloped wet savannas available for agricultural expansion.

In the case of solid biomass, woody and stalk-like biomasses such as energy forest wood and wood from short-rotation plantations (e.g. poplar or willow) and energy grasses (e.g. miscanthus) are suitable raw materials for cultivation. Woody and stalk-like biomasses are also available as waste feedstocks: Wood residues from forestry (e.g. forest residues), from wood processing (e.g. sawmill residues) or after use (e.g. waste wood), and agricultural and landscaping residues (e.g. straw, roadside greenery, landscaping greenery). The components of the solid biomass are decisive for the conversion, especially cellulose, hemicellulose and lignin. The specific energy density of solid biomass is naturally low, i.e. the harvesting and collection of biomass is time-consuming and requires large transport volumes and corresponding (temporary) storage capacities. Storage takes place in the open or in halls. In special cases, silo storage (e.g. sawdust, cellulose residues) may be necessary. The "CAAFI Feedstock Readiness Level" describes the technological and commercial maturity of the raw material. The production of lignocellulosic feedstocks for energy purposes corresponds to CAAFI Feedstock Readiness Level 5 scaled commercial production system validation and scaled commercial testing. The classification means that the conversion pathway as such is proven to work and is in the process of being scaled up. This indirectly implies that potential risks of feedstock production on an industrial scale are not yet known or cannot be completely excluded, even if the associated conversion process has a higher readiness level.

In contrast to solid biomass, vegetable oils from oleaginous crops, oleaginous green plants or algae can be produced by mechanical pressing and/or solvent extraction, provided that the oils are long-chain triglycerides. Suitable oil plants are those intensively cultivated for the food industry, such as sunflower species cultivated for the food industry, such as sunflower oil, soybean oil or palm oil, and the vegetable oils that are not in competition with food and animal feed, Gold of Pleasure (Camelina sativa; annual) and Jatropha (Jatropha curcas, perennial). Due to its high fatty acid content, rapeseed is not used as a feedstock for SAF. Vegetable oils are produced either by mechanical pressing of the oleaginous fruit or by solvent extraction. In the case of jatropha oil, which is extracted from jatropha seeds and, due to its toxicity, is not in direct competition with the food supply, water-based extraction offers the possibility of using the jatropha meal as animal feed protein after detoxification. Extraction processes are also used for other annual oil crops, such as canola (Camelina sativa). Mechanical cold pressing with a screw press is an alternative or pre-treatment stage to oil extraction.

Jatropha curcas as a perennial crop requires a growing period of 3–4 years after planting the seedlings under subtropical climatic conditions. This cultivation method inevitably leads to an economically complex investment model. Land acquisition, land development, field preparation, planting and a subsequent 3–4 years of labour-intensive field management without significant oil yield results in a high capital commitment due to unavoidable start-up losses, which, depending on the size of the farm and the selling price, may take 7–8 years to recoup (break-even). Over the life of the investment, a return on capital at generally accepted industry levels is achievable, but this requires stable crop yields after the end of the growth phase. Growing Jatropha on degraded arable land is feasible, but results in lower oil yields and a lower return on investment, or would force sales prices to break even, which is not sustainable in the market. The CAAFI Feedstock Readiness Level for Jatropha is Level 6 – Commencement of Full-Scale Production. Due to a number of unsuccessful Jatropha curcas planting projects in Kenya, Tanzania, Mozambique and Indonesia in 2010–2015 for various reasons, research on seed improvements as well as further investment in farm projects were discontinued (Research Gate, 2017).

Camelina sativa is grown in temperate climates as an annual field crop, with planting in late fall and harvesting in winter, or planting in spring with harvesting in late summer. As a crop, Camelina sativa can only be grown in a 3-year rotation to avoid soil degradation. The crop is harvested mechanically by combine harvesters, which separate the oilseed from the stalk crop. The press cake can be used as a protein source in animal feed production, generating additional income. Field rotation with other cereals such as wheat and maize, as well as field harvests of all crops concentrated on a few days of the year, leads to high investments in storage silos and warehouses, as well as in agricultural equipment (tractors, combines, transport trailers, etc.). From an economic point of view, the cultivation of Camelina sativa can only be calculated in a mixed calculation with the cereals to be grown in rotation. Combined business plans for an integrated farm model with Camelina sativa and two other cereals also show high start-up losses for the above investments, which reach break-even after 7–8 years, with total returns over a 20-year period at industry level before taxes, based on current market prices for vegetable oils and cereals. For the use of vegetable oils as an energy feedstock, however, only the yield of the oil is relevant. Regardless of the process used to extract the oil, solids and impurities must be removed by filtration. The native oil is stored as crude oil in tank containers for further transport to oil processing plants. The CAAFI Feedstock Readiness Level for Camelina sativa corresponds to Level 8 – Commercialisation. This reflects the level of maturity of "industrially scaled production".

7.1.2 Feedstock Logistics

For the transportation of feedstocks, a distinction must first be made between cultivated biomass and residual and waste materials.

7.1.2.1 Cultivated Biomass

The harvested material is transported from different cultivation areas to a collection point. This may be a temporary storage facility or a pre-treatment facility. This point is referred to in the LCA as the "first gathering point" because it is where the harvested material from the different growing areas is collected. For example, an oil mill serves as a pre-treatment facility where oil fruits are separated from their shells, and the oil seeds are then mechanically pressed or extracted. The hulls and press cake remain on site for use as fertiliser, biogas or organic fuel. The processed raw material is then transported to the SAF production facility in suitable containers (truck, rail, barge or pipeline). At the SAF production site, the raw material is either temporarily stored or further pre-treated, e.g. hydrogenated to break up long molecular chains. The SAF is then produced and transported to a blending depot, which may be located on the same site. After blending, the JET A-1/SAF is transported to the airport tank farm for storage. Transportation options include truck, rail, barge, ocean-going vessel and pipeline. Once on the airport tarmac, it is transported to the aircraft by bowser or via an underground hydrant pipeline network.

Depending on the SAF production process, raw material costs can account for up to 65% of manufacturing costs. The energy content of the raw material influences the transportation costs: Cultivated biomass such as straw and corn as well as wood from short-rotation plantations require a large transport volume with a low energy content. Only after pre-treatment – as is done with vegetable oils – does the transport volume decrease while the specific energy content increases. Vegetable oils are easy to store in tank containers and therefore inexpensive to handle. Cellulosic biomass has a very low energy content and requires very high transportation costs due to its volume. This results in a classic arbitrage dilemma: large-volume biomass can only be produced in a limited area around the SAF production plant, because the transportation costs affect the economics of the SAF production process as the distance increases. The SAF plant is therefore dependent on the surrounding area for its economic life. However, price arbitrage means that the raw material producers set the selling price as high as the most distant raw material producer can still economically supply the SAF plant. The costs of the nearest alternative thus determine the prices of their own sales. This form of profit skimming tends to result in raw material prices at the expense of the profit margin of SAF production. This dependency does not exist with vegetable oils because the raw material can be transported easily and over long distances at reasonable cost.

7.1.2.2   Residual and Waste Materials

Residues and wastes are treated in the same way as cultivated biomass. For example, wood chips, forest residues and sawmill waste can be classified as volume shipments. However, the collection of these residues is costly and the individual quantities are comparatively small. The same applies to used cooking oil (UCO), which is collected in tank containers from the food processing and catering industries. Municipal solid waste (MSW) collection in cities and metropolitan areas is also small-scale, but more complex. The volume of waste is limited to the capacity of one collection truck. Since metals and non-combustible or gasifiable materials – such as glass – must be separated from the waste before use, the waste transfer stations are the "first collection point" for this raw material. The waste is then transported in bundled containers to the SAF production facility, where further pre-treatment can take place. UCO and MSW enable stable supply chains, as the raw material accumulates regularly and in approximately equal quantities. This means that long-term transportation contracts can be concluded and there is competition among transportation companies due to the transportability of the raw materials.

Consequently, the supply of raw materials requires not only a consideration of the availability of raw materials but also a comprehensive location analysis for SAF's production plant and a simulation of the transportation costs and routes for various alternative locations. The long-term availability of railroads and, in the case of truck transportation, the passability of roads in winter must be taken into account. Possible bottlenecks, especially bridges and tunnels, should also be examined, and their technical condition and availability should be included in the concept planning. At least one alternative supply and delivery route should be calculated as a worst-case scenario before the site is selected.

7.2   NON-BIOLOGICAL FEEDSTOCKS

Electrofuels, known in aviation as eSAF, are based on non-biological feedstocks such as electricity, hydrogen and natural gas. The EU regulation also requires a minimum of 70% CO_2 emission reduction as a threshold for EU-ETS accounting. In particular, the EU Commission has issued the so-called "Delegated Acts" clarifying Article 28 of the EU-RED II Directive, which specifically define and explain how the emission reduction potential of RFNBOs is to be assessed.

In order to meet the EU requirements for the use of non-biological feedstocks, a number of conditions must be met:

1. Electricity must be generated from solar fields, wind farms or hydroelectric power. The use of such power must be in time with the operation of the electrolyser, which separates hydrogen and oxygen from water. The background is the requirement of "additionality" of the green electricity used. This power must be generated by dedicated facilities in order to avoid conflicts with existing users of green power, who might otherwise suffer from supply shortages.

2. Electrolysis and PtL production require large quantities of water, which are initially taken from the municipal water network. However, given the availability of potable water, much of the process water can be treated and reused. As a "green" fuel, disproportionate water consumption would be inappropriate and would support the argument of some NGOs that industrial water consumption is increasingly at the expense of groundwater availability. Water treatment therefore underscores the claim of climate-friendly fuels to an intact environment.

3. In addition, PtL kerosene requires a carbon source because the end product will be a liquid hydrocarbon for aviation. There are three alternatives for providing CO_2 as a feedstock:

A. CO_2 is extracted from the flue gas stream of a CO_2 point source and fed to the PtL plant. Large CO_2 emitters are steel and cement plants, whose flue gases contain a high proportion of CO_2. The close proximity of the PtL plant is an advantage for this alternative. The CO_2 separated from the flue gas can be fed into the eSAF production process at low cost, which has a positive effect on the overall cost of eSAF production in view of the high cost of electrolysis. From an environmental point of view, this alternative is rejected by the NGOs, as the connection to an industrial plant requires a permanent CO_2 supply and thus ensures the long-term existence of the fossil-fuelled industrial plant, which is not desirable from an environmental point of view. However, the counter argument from the eSAF producers is that the recycling of exhaust CO_2 displaces the introduction of new CO_2 from petroleum-derived jet fuel, resulting in a net reduction of CO_2 entering the atmosphere. In recognition of the prohibitively high production costs for eSAF, the EU has set a deadline in its Delegated Acts to the EU-RED II/RED III, according to which industrial point sources are permitted as a carbon source for eSAF production until 2041 (EU 2023/1185). However, taking into account the usual depreciation period for industrial plants, which is 20 years, the current EU regulation leads to an investment problem, because if the construction of a PtL plant is completed in 2026, the framework conditions for operation will only apply for 14 years, and the owner would still have to invest in another carbon source for the remaining useful life. Theoretically, the depreciation period for such a production facility could be set at a shorter period, provided this is recognised by the tax authorities for income tax purposes. From a commercial point of view, a shorter depreciation period leads to higher operating costs and thus to higher manufacturing costs for eSAF, which hampers the market introduction of this promising technology through higher prices.

B. CO_2 is extracted from biomethane gas, which has the same chemical formula (CH_4) as natural gas. Biomethane gas is produced by the anaerobic fermentation of plant residues and manure in biogas plants. The most common use of the gas today is to generate electricity in combined heat and power plants by burning it in an engine, and to use the excess heat to heat swimming pools, greenhouses or other heat consumers.

The digestate from the biogas plant can be used as a high-quality fertiliser in agriculture. As a feedstock for PtL jet fuel production, the biomethane is either fed directly into the plant or fed into the local natural gas grid via a product exchange and withdrawn as natural gas at the eSAF production site (banking system). The emission reduction certificates are then transferred to the operator of the eSAF production, as long as the withdrawn quantity is identical to the supplied quantity. Chemically, the two products are identical; only the emission reduction certificates are transferred. This makes eSAF an environmentally friendly fuel, but it contradicts the definition of the RFNBOs, according to which eSAF may not contain organic raw materials. This dilemma can only be resolved to a limited extent, for example by classifying the volume of biomethane in the end product as "advanced biofuel" and the remaining volume as RFNBO. From the producer's point of view, the use of biomethane is an environmentally friendly and economically viable solution, which enables safe plant operation for the upcoming first generation of plants.

C. The "gold standard" is CO_2 supply by extraction from ambient air. In this process, the CO_2 contained in the ambient air is filtered out and fed into the process as a feedstock. However, the technology for extracting CO_2 from the air is still in the early stages of research and development, with a current TRL of 4. The current operating costs for Direct Air Capture (DAC), as the process is called, are very high at around USD 540/tCO_2 (Block, Simon et al., 2022). Industrial application of the technology is not yet available and, due to the technical requirements, the resource consumption is much higher than the environmental benefit. Nevertheless, environmental organisations promote DAC as the only tolerable source of CO_2 for eSAF production. In terms of content, however, the question arises as to why CO_2 from flue gases is unacceptable despite the fact that it would end up in the atmosphere only to be filtered out again at great expense. From an environmental and economic standpoint, the technology must first be matured to the point where its use becomes ecologically sound and economically feasible.

7.3 SOCIAL AND ECONOMIC UNCERTAINTIES OF FEEDSTOCK PRODUCTION

The availability of raw materials for the production of SAFs is becoming increasingly important, as special requirements have to be taken into account for the use of biomass as an energy feedstock.

1. First, it must be ensured that only certain biomasses are suitable and approved for production, depending on the SAF production process.
2. Next, it is necessary that the cultivation of this biomass is carried out professionally, that the agricultural land is suitable for the specific biomass, and that the cultivation region has sufficient rainfall and that the distribution of rainfall over the year is sufficient for a good harvest.

3. The employees of the cultivation company must be professionally qualified so that the cultivation of the area is carried out professionally.
4. Plants approved for energy cultivation are suitable for inclusion in emissions trading systems and can therefore also be certified.
5. For the cultivation to be economically successful, suitable seeds must be used to achieve the predicted crop yields.
6. The cultivation region should have a suitable infrastructure to transport the biomass for further processing.

In terms of cultivation regions, farms in the US, Canada and Central Europe are not considered critical in terms of the above criteria. In order to expand the production of SAF based on the use of biomass, new cultivation areas will have to be developed in Central and Latin America, Africa and Australia, and the necessary infrastructure will have to be built in advance. There is sufficient land potential in these regions that does not compete with food production. The question is whether sub-Saharan Africa and countries such as Paraguay and Uruguay should be considered for energy crops.

Biodiversity, i.e. the preservation of species diversity, is the most important consideration in the selection of areas. Although industrialised countries have no problem with large-scale monocultures, the development of new cultivation areas must comply with rules that support biodiversity.

LUC (Land Use Change) refers to the change of use of an existing cultivated area, for example from food production to energy crops. ILUC (indirect Land Use Change) defines the consequence of the displacement effect when previously untouched land is subsequently converted into new cropland to replace land lost to energy crop cultivation. Soil quality is a key determinant of crop yield. For energy crops, it is sufficient if the soil quality is average and sufficient to feed the plants. Trials of the perennial Jatropha curcas as an oil crop have been largely unsuccessful, either because they were grown on degraded soils or because the seeds were wild rather than cultivated. Field rotation is required for annual crops, because growing energy crops without field rotation leads to soil degradation and thus reduced yields.

Due to the unstable political situation in various countries and the lack of transportation infrastructure, major agricultural investments have not been made in developing countries, even though the development of professional agriculture in these countries would significantly reduce hunger and poverty. However, as long as farms in the US, Canada, Ukraine and Russia can offer grain and vegetable oils at lower prices than would be the case with production in a developing country, such investments will not be made. Instead, the EU Commission in particular is relying on eSAF, even though production costs are significantly higher than those for SAF in Africa or Latin America. Assuming a market price for SAF from HEFA or ATJ, such a price is also used as a basis for production in developing countries. Due to the higher production costs, this reduces the profit margin, although a higher profit margin would be necessary as a risk correction when considering the country risks of developing countries.

Assessment criteria for agricultural labour conditions are provided by the UN Compact (UN, 2023) and the Food and Agriculture Organisation of the United Nations (FAO). To this end, the FAO has developed a catalogue of criteria for assessing food security (FAO, 2013) and a toolbox for determining the level of food security in the case of simultaneous cultivation of bioenergy feedstocks (FAO, 2010).

7.4   FEEDSTOCK CERTIFICATION

As part of the life cycle analysis, the sustainability of biomass cultivation and the emission reduction effect are the essential components of biomass certification.

The EU-RED II Directive limits the use of biomass in the transport sector to 14% of the final energy value.

In order for renewable fuels, including SAF, to be included in the EU-ETS, an emission reduction of at least 70% compared to the fossil alternative is a prerequisite for inclusion in emissions trading. The achievable emission reduction is documented through the certification of the entire process chain by a certification system. The two preferred certification systems in Europe are ISCC (based in Cologne, Germany) and RSB (based in Geneva, Switzerland). In addition, there are 6 other systems accredited in the EU. The life cycle analysis starts with a 100% emission reduction on the cultivated area. The initial value is reduced by offsetting all emission-increasing factors, such as the use of artificial fertilisers to increase yields. In addition, all transportation and processing emissions to the airport depot are subtracted. The remaining emission reduction value must be at least 70%. If this is not the case, all emissions can be recorded individually, or default values can be used.

If the effective emission reduction is less than 70%, a decision must be made whether to forgo emissions trading credits (and buy allowances instead) or whether the minimum reduction value can be achieved through targeted measures. These include, for example, investing in solar and wind energy to supply the farm with green electricity, increasing the use of organic fertilisers, and optimising transport routes and means of transport.

The EU-RED II Directive limits the use of biomass from food and feed crops in the transport sector to 7% of the final energy value to avoid potential ILUC conflicts.

7.5   FEEDSTOCK STORAGE AND AVAILABILITY

The biomass-based SAF pathways depend on the availability of sufficient quantities of raw materials and thus on the harvest yield and harvest timing.

Depending on the weather conditions, the prevailing temperatures and the amount of precipitation, the yield of an area varies from year to year. However, agricultural operating costs are largely independent of crop yields. Prices for agricultural products are set in the marketplace based on the size of the harvest. In contrast to crude oil production, harvest volumes can vary by up to 30% from year to year, with a corresponding impact on commodity prices. As a result, the supply of biomass-based feedstocks to the SAF market is subject to price fluctuations similar to those in the oil market, but in this case caused by the available harvest volume rather than an expected supply shortage on the refinery side. By storing the crop, SAF feedstocks can be held for several weeks, thereby dampening supply fluctuations. Typically, the crop is temporarily stored in elevators or warehouses. In mechanical grain harvesting of annual crops, the husks and stalks are separated by the combine harvester during the harvesting process in the field. This process is also used for annual oil crops.

Annual crops in cereals can be harvested twice a year (winter and summer), which has a positive effect on storage time and availability of the raw material. Annual oil crops are harvested once a year. However, there is the possibility of short-term intercropping. Utilisation of harvesting machinery and storage facilities is therefore concentrated on the respective harvest periods.

A continuous supply of raw materials is essential for SAF's production. At the same time, the storage capacity of organic biomass is limited in time. Oil plants in particular lose their oil content with increasing storage time. As SAF production increases, this dilemma will become another factor in evaluating the security of supply of scaled SAF production facilities. SAF producers therefore need to develop a diversified supply strategy to ensure continuous plant utilisation with stored raw materials. It is important to consider whether the feedstocks are US EPA certified and therefore processed into SAF or synthetic diesel to be sold in the US market, or whether they are EU-RED II (future RED III) compliant feedstocks whose end products are to be sold in the EU27.

The segregation of feedstocks at the SAF production site means that the production plant must sequentially determine for each product batch which market it is producing for and which type of feedstock it is using. For international SAF producers, this increases complexity and production costs.

The palm tree is a perennial source of raw material for SAF. The fruit bunches are manually cut from the tree with sickles and collected. The fruit bunches have a diameter of up to 80 cm and can weigh up to 30 kg. The fruit bunches are transported by truck to the oil mills where they are cleaned and then threshed. The palm oil remains in the processing plant for further processing to obtain edible oil. The palm kernels also contain oil, but it is not suitable for human consumption. The kernels are mechanically pressed and the resulting paste is mixed with water and heated to extract the palm kernel oil. As an edible oil, palm oil is found in many foods, such as frying fat, margarine or chocolate, and as animal feed in chicken fattening. The use of palm oil for energy purposes has been almost completely phased out after many NGO protests and is now limited in the EU27 to plantations that existed before 2020 (EU Regulation on deforestation-free products, 23 May 2023).

The Roundtable for Sustainable Palm Oil (RSPO) is considered a trustworthy certification system for palm kernel oil. The basic problem with using perennial crops is the low level of automation in harvesting. In terms of cost accounting, the cultivation of annual plants is more economical and the break-even point of an investment in the cultivated area is reached within 3 years at the latest. Therefore, from an investor's point of view, mechanised agriculture with high machine use is the preferred investment option. In addition, pests and diseases can cause lasting damage to perennial crops and thus jeopardise the long-term success of the harvest. For SAF production, the classic four-field rotation is therefore the cultivation method of choice for energy crops.

The demand for energy feedstocks is also affected by the competition between SAF and synthetic diesel. While classic first-generation biodiesel (FAME – Fatty Acid Methyl Ester) is mainly used as a blending component due to engine technology (e.g. 7% B7 blend in Europe) and is only used as a pure fuel in robust engines in agriculture, the demand for synthetic diesel is steadily increasing. This is also driven by stricter carbon intensity requirements, such as the LCFS in the US state of

California. Synthetic diesel is produced in two specific products: HVO (hydrotreated vegetable oils) and HEFA (hydrotreated esters and fatty acids). Both production processes use organically grown biomass. HEFA also utilises animal fats and slaughterhouse waste. HEFA refineries also produce diesel and SAF as by-products, so both products are in direct margin competition.

The long-term availability of biogenic feedstocks for SAF therefore implies restrictions in the utilisation of these feedstocks for diesel production. The current SAF production of approximately 0.5–1.0 million tons in 2023 does not yet lead to predatory competition, but the combined demand for synthetic diesel and SAF will lead to a resource bottleneck in the near future if the aviation industry – especially in the US and Canada – rapidly increases SAF blending quotas. A shortage of supply will inevitably lead to higher market prices for agricultural commodities.

Small farmers and small cooperatives are found primarily in the Far East, where large farms account for only a small portion of agricultural production. For this group of producers, there is the position of the "middle man" who stands between the farmer as producer and the buyer of the products. The middleman negotiates the contracts and acts as a representative of the farmers he represents to the buyer. For his work, the middleman receives a fee, such as a percentage of the sales revenue. If market prices change, there is a risk that the middleman will sell the production to other buyers at higher prices in order to receive a higher commission. This is contrary to the requirements of contractual obligations and loyalty, but can be observed in daily practice in some countries. The buyer who has been defrauded in this way is then faced with the task of finding a short-term replacement for the missing crop. Similar business practices may occur when a crop has already been completely sold by the group of "middlemen". Of course, small farmers can also be involved in the cultivation of energy crops in order to supply SAF with raw materials. For this purpose, fertiliser is also supplied to the participating farmers, with the buyer deducting its procurement costs from the remuneration for the raw materials supplied. Unfortunately, some smallholders see an opportunity to sell the fertiliser instead of applying it to their fields to increase crop yields. If these fertilisers are missing, the harvest will be lower than expected, which means that the buyer will not be able to meet his contractual quantities and the farmer will receive a significantly lower income, from which the cost of providing the fertilisers has to be deducted. For the sake of completeness, it should be mentioned that the majority of farmers and middlemen are contractually loyal and reliably fulfil their contracts. However, the high growth of air traffic in Asia means that in the future many farmers will have to be involved in the production of raw materials in order to provide the necessary quantities of raw materials for SAF production. The production of SAF will therefore create many new jobs, especially in emerging and developing countries, which will benefit economic development in rural areas. However, in order to ensure the reliability of the feedstock supply chain, it is essential to examine the feedstock business model for alternatives.

As waste gasification followed by Fischer–Tropsch synthesis comes to market, a new class of feedstock is being added to the feedstock supply chain: MSW, which is currently incinerated or landfilled. However, waste pre-treatment is costly and waste gasification requires a minimum energy content in the residual waste; otherwise, the gasification process produces only marginal amounts of syngas. An argument in favour of waste gasification is that it solves a waste problem in metropolitan areas and eliminates the need for landfills. It may even make sense to reopen former landfills to use the waste for energy recovery.

Furthermore, the commercialisation of PtL jet fuel with the provision of hydrogen offers unlimited feedstock availability in the future, provided that sufficient "green" electricity can be made available.

In summary, feedstock availability is directly related to feedstock prices. With steadily increasing demand, all SAF technologies will find their market share as long as feedstock availability and feedstock prices do not limit production.

7.6   PRE-TREATMENT OF FEEDSTOCKS

SAF production is focused on the operability and scalability of the production facilities for the manufacture of SAF. In co-production, naphtha (light gasoline), diesel and SAF are produced from the raw materials used. Since most of the production processes are new, interest is focused on the further development of the first generation of plants and the stability of the production conditions (first mover problem). However, the fact that significant investments are also required in the upstream supply chain to stabilise the entire supply chain is often overlooked. In contrast to the process chain in crude oil processing, there is a lack of supply chain integration and central control of the individual processes in a form that is coordinated with the production target to be achieved. In order to achieve the logistics competence of the petroleum industry, which has been developed over a period of 75 years, a high level of performance is required from all participants in the SAF process chain and a willingness to provide "on-time" services that are seamlessly integrated. The history of the industry shows that such system integration requires quasi-military logistics and production resources along the process chain in order to come as close as possible to the efficiency level of the oil industry.

Investments in the feedstock supply chain repeat the sequence that can be observed in attracting investors for new SAF plants: (1) suitable processes and means of production are selected, (2) investors are approached to finance the necessary procurement and construction measures, (3) investors expect revenues to be secured through long-term supply contracts, (4) project developers for new SAF plants can only make these commitments if they also receive corresponding legally binding commitments from the buyers of their products. This chain reaction is not a problem if the companies in the supply chain have completed the necessary preliminary work and procurements by the time the SAF plant starts production.

With regard to the additional investment requirements for the supply of raw materials to SAF, it must be examined which existing investments in the supply of raw materials can continue to be used and merely be supplied to a different recipient for SAF production. In the case of an expansion of production in terms of volume, additional investments must be made in known technologies and products in this segment. These include tractors, transport vehicles, harvesters, warehouses and elevators, as well as railway trains and possibly inland waterway vessels.

In the pre-treatment of woody biomass, torrefaction is a common technology used to make the subsequent gasification process more efficient. Torrefied wood is similar to charcoal and leads to faster chemical reactions in the gasifier.

Additional investment in this pre-treatment technology may be required if the gasification technology includes the use of wood chips and wood from short rotation plantations (as feedstocks for ASTM D 7566 Annex 1 FT-SPK – Fischer–Tropsch – Synthetic Paraffinic Kerosene). Waste sorting plays an important role in the utilisation of residual materials. Metals can already be largely extracted from the waste sorting belts using magnets. Sorting glass from household waste is more complex. The glass remaining in the waste is melted and sintered in the plasma gasification process required for waste gasification with reactor temperatures of up to 1,700°C. Ideally, the ash content increases as the chemical molecules break down and is discharged from the reactor as hazardous waste and then landfilled. Gasification of toxic substances in residual waste – such as PCBs or PCPP in plastics, which release dioxins and furans – requires precise analysis to determine whether these pollutant molecules are completely broken down.

Hydrogen production as a precursor to PtL production requires green electricity, which must be provided by new solar or wind power plants. In addition, it has to be checked whether the existing grid capacity is sufficient or whether additional grid extensions are required. For a 200 MW electrolysis plant, an investment of approximately EUR 1.2 million must be calculated. The total investment is USD 240 million. For a wind turbine with a capacity of 4 MW, an investment of approximately 7.6 million EUR is required for the turbine, foundation and installation. For a wind farm with a capacity of 200 MW, the cumulative investment amount is 380 million EUR. In the investment analysis, the necessary follow-up investments in the supply chain should therefore be carefully assessed and calculated for the overall financing of an SAF project. The success of the project is jeopardised if the supply chain is incomplete or does not function for other reasons.

Conclusion: The selection of topics in this section partly shows the complexity of the SAF supply chain, which has to be considered in the course of project planning for new SAF projects. So far, the topic of SAF production has been limited to North America and Europe. The Mercosur countries, due to their geographic location in South America and labour and skills potential, will become offshore production locations for Europe in the coming years, before national governments also issue SAF mandates for their air traffic. Aviation as a global mode of transportation requires global SAF supply solutions. In the coming years, many flights will take off from Europe in a relatively climate-friendly manner and land in Europe in a climate-unfriendly manner on the return flight. CORSIA will do little to change this, as the majority of the fuel used in international traffic will not be offset. In the US, domestic traffic is dominant, so the problem is less significant in North America.

As long as blending rates remain in the single-digit range, the oil industry will be able to make up any shortfall with fossil JET A-1 if, for example, an SAF producer reports a prolonged technical disruption or the company has to file for bankruptcy. This means that the industry will be in a situation where the legal quotas are not met, but no aircraft will be grounded as a result. From a legal point of view, it needs to be clarified whether an unforeseen production stoppage – for whatever reason – is to be considered a force majeure event and whether SAF's replacement cannot be guaranteed in case of a tight market supply. This would make an EU penalty payment for failure to meet quotas inappropriate. Consideration of the SAF supply chain is therefore an essential key factor for the market success of SAF production.

8 Production of SAF

The technical development history of SAF's production goes back to the German chemists Franz Fischer and Hans Tropsch, who in 1925 patented the process they had developed for producing synthetic hydrocarbons from carbon monoxide and hydrogen derived from the gasification of coal ("FT synthesis"). In the 1960s, an oil import embargo was imposed on South African industry because the government in the capital, Pretoria, maintained racial segregation ("apartheid") and openly discriminated against members of certain ethnic groups. To meet the demand for fuel, the national oil company, SASOL, produced synthetic diesel and jet fuel from black coal, supplying both domestic and international aviation. For years, many international airlines probably flew from Johannesburg and Cape Town back to their home countries on FT synthetic jet fuel without any problems.

Based on FT synthesis, Shell and SASOL, together with the FAA and ASTM International, developed a qualification concept for alternative aviation fuels in the 2000s, the formal procedure for which has since been standardised with the ASTM D4054 standard in order to standardise processes and ensure a level playing field for all potential producers in the application, testing and approval process.

ASTM International (formerly ASTM American Society for Testing and Materials) is a consensus-based standards organisation in which the relevant stakeholders in a field work together to develop standards and specifications to address industry needs. The consensus-based approach employed by ASTM adequately examines all aspects of the subject and can satisfactorily answer all questions with its panel of experts. For aviation uses, the safety and performance of sustainable aviation fuel is the most important issue of all (Gurhan et al., 2021). The Aviation Fuel Subcommittee D02.J0, chaired by Mark Rumizen, formerly the FAA's Senior Technical Specialist for aviation fuel, represents all stakeholders: oil companies, SAF producers, engine manufacturers, airframe manufacturers, airlines, airports, regulators such as the US FAA and the EU EASA, and scientists from universities and research institutes. The committee focuses exclusively on technical and operational issues related to the production, storage, distribution, refuelling and safe use of aviation fuels, even under challenging conditions. Membership in ASTM Fuel Subcommittee D02J0 is voluntary and open to any company or person interested in participating in the process. Issuance of new standards and changes to existing standards are accomplished by voting of the subcommittee members, and all unresolved issues are deliberated until there is a unanimous – and affirmative – vote of the full subcommittee. The question of whether a new production process is safe and whether the resulting fuel can be used in an aircraft is not a question of majority decisions, but must be answered positively from all perspectives. This has been a complex experience in the first approvals after 2009 and has sometimes required patience on the part of the applicants, but in retrospect, it was the right approach. On the one hand, no party to the process wants to approve a fuel unless they are fully convinced of its safety; on the other hand, every applicant must also be aware that an incident or accident that can be traced back to their fuel type will result in the economic bankruptcy of the company along with collateral consequences to the entire SAF industry. The ASTM Fuel Subcommittee has therefore acted in accordance with the system and enforced the topic of reliability and safety as a priority for all approval decisions – sometimes even against uncomfortable questions (Rumizen, 2018). The most important statement for SAF is that all passengers and aircraft crews can fly safely with SAF blends because the underlying safety concept is consistently and fully implemented.

Today, the qualification of a new SAF fuel type requires, among other things, the production of fuel test quantities that can range up to approx. 900,000 L or approx. 240,000 USG if component rig testing and engine testing is required. However, smaller fuel quantities of under 100 gallons have been successfully used to support qualification based solely on fuel property and compositional testing. The qualification process, if the use of the engine test rigs of the engine OEM is necessary, may require expenditures of up to USD 5.5 million. The results of the engine manufacturers' technical investigations are presented and explained to the Fuel Subcommittee. The duration of a qualification process ranges from 24 to 36 months, depending on the availability of engine test beds for the test procedures and the complexity of the qualification requirements. The FAA reviews all documentation and test results prior to a final vote. In 2020, an option for "fast track" qualification under certain conditions was introduced. The fast-track procedure is limited to SAF with conventional hydrocarbon compositions and a blending limit of 10% is imposed in the final specification when issued. This will at least shorten the time to market and allow an accelerated return on investment (US Department of Energy, 2020; Rumizen, 2021).

As mentioned above, JET A-1 contains aromatics as part of the manufacturing process and composition of the jet fuel. The aromatics play an important role in the fuel lines and aircraft tanks because they prevent the rubber or plastic seals in an aircraft's fuel system from becoming brittle, thus preventing uncontrolled leakage of jet fuel into the aircraft fuselage. Because aromatics in jet fuel also have undesirable effects, such as soot formation and in the durability of combustion components, airframe and engine manufacturers are now working on new materials for seals that will be able to tolerate non-aromatic jet fuel without leaking into the fuselage. However, there are approximately 22,000 existing aircraft in service that rely on aromatics in jet fuel. It is therefore necessary to determine which types of aircraft can have their seals changed, how much this will cost and whether the change can be made during a routine layover or whether a separate layover of the aircraft is required, which would result in high modification costs due to loss of revenue.

As a JET A-1 compatible aviation fuel, non-aromatic SAF is subject to another limitation: the ASTM specification specifies a density range of 0.775 kg/L and a maximum density of 0.84 kg/L for JET A-1. The commonly used standard density for JET A-1 is 0.80 kg/L. The cockpit instruments in the aircraft as well as the computer programs that calculate the fuel requirements of a flight prior to a flight, taking into account the legal minimum quantities and the planned flight route as well as the weather situation en route and at the destination airport, assume the standard density of 0.80 kg/L, unless the airport fuel depot reports a significantly lower density in the storage tanks, which must then be taken into account in the quantity calculation.

Non-aromatic SAF typically will have a density below the JET A-1 minimum, and the consequence of a lower fuel density is a corresponding increase in consumption during the flight to compensate for the lower energy content of the jet fuel. As a result, the minimum fuel levels may be exceeded before reaching the destination airport, and the remaining fuel may not be sufficient to fly to an alternate airport due to weather conditions. Fuel density is therefore a key flight safety parameter.

In six of the seven qualified SAF technology paths, the SAF types do not achieve the required minimum density of 0.775 kg/L, resulting in significant additional consumption that would not be reflected in the calculation of a flight's fuel quantity and would therefore lead to operational problems in flight operations. For this reason, the ASTM Fuel Subcommittee has currently restricted the use of SAF and only approved it as a "Synthetic Blending Component" (SBC) with a maximum blend of 50% to JET A-1. Individual processes have been approved for as little as 10% blending.

Only the "Catalytic Hydrothermolysis Jet" (CHJ) production process meets the JET A-1 density requirements and also contains aromatics. However, the ASTM Fuel Subcommittee initially limited this process to a 50% blend. The combination of SAF and JET A-1 ensures that the blend is within the jet fuel specification ASTM D1655, thus meeting all the parameters for the fuel and making the blend equivalent in composition to fossil JET A-1. In view of the limited production volumes of SAF producers, this restriction does not hinder the further development of SAF blends, as the blending mandates (in the EU) are in the single-digit percentage range until 2030.

However, it must be pointed out to SAF users who want to use quite high blends for individual production batches that the fuel properties of the respective JET A-1 batch determine the maximum blending rate and that blends above 30% may already be in the limit range of individual parameters that do not allow a further increase in the blending rate. Therefore, each fossil batch must first be tested for its parameters before a decision can be made on the SAF blending rate. This makes the blending process complex, but primarily serves the purpose of flight safety, even though no surprises are to be expected with blending in the single-digit percentage range. Nevertheless, care is paramount and carelessness jeopardises flight safety.

With reference to the first qualification of FT-SPK as a blending component in 2009, natural gas and biomass were also approved as raw materials in addition to coal. The SAF produced in this way can only generate US RINs or be counted in the EU-ETS if it is biomass-based (BtL – biomass-to-liquid) or, in a cascade use, if it is WtL (waste-to-liquid) from waste that is classified as emission neutral in the second use stage. In this way, SASOL's coal-based synthetic fuels in South Africa were retroactively rehabilitated and the research achievements of SASOL, which successfully established FT-SPK as a synthetic fuel in South Africa several decades ago, were recognised.

With the first ASTM approval came another industry problem: There was a lack of consistent and unambiguous terminology to describe the various products. The first common term was "aviation biofuels" as a generic term. This was replaced by "Bio-SPK" until the current term "SAF" was finally established in 2018. The first ASTM approval was called "BtL" for many years because it was assumed that the feedstock would be biomass. However, after no investors could be found for this technology, it was switched to natural gas as the feedstock and this process route was called "GtL" – Gas to Liquid, which is now being used on a large industrial scale by Shell at the Pearl project in Qatar, using natural gas as the feedstock for synthetic fuel production.

Continuing the FT-SPK story, it was finally the US company Fulcrum that was able to start up the first waste-based FT-SPK plant and produce SAF in 2022, while Red Rock Biofuels in Lakeview, Oregon, USA, had to file for bankruptcy in early 2023 with its BtL refinery still under construction, after Choren Industries' BtL plant, completed in 2014 in Freiberg near Dresden, Germany, also had to file for bankruptcy due to technical problems (Erneuerbare Energien, 2011). The Solena-British Airways "Green Sky London" project also failed to progress beyond the planning stage due to a lack of funding (Biofuel International, 2023; Business Travel News, 2016, Renewable Energies, 2011). These three examples illustrate that new technologies that work excellently on a small scale may well run into existential problems when scaled up to industrial plant sizes – either because of technical problems or because of financing problems when investors have doubts about the successful implementation of the concept (Figure 8.1).

The so-called "valley of death" describes the funding gap for a new technology that is typical for start-ups. The initial seed capital is used up in research and development. Cash flow becomes negative and fresh capital is needed. The product launch brings the new technology to market. Cash continues to be required for project funding until the commercial success of the technology or product launch improves the cash position. Once the baseline is reached, the start-up is considered a successful business.

With regard to SAF's new technologies, the "valley of death" situation is exactly what is happening. Process development must be advanced to the point where stable production can be demonstrated in a pilot plant. Otherwise, the volumes required for ASTM approval cannot be produced. With approval costs of $7.6 million, initial seed capital of $10–12 million is required to get the process approved. For the next

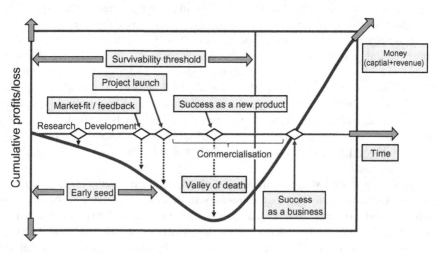

FIGURE 8.1   The Valley of Death for Start-ups.

(Source: Osawa and Miyazaki, 2006) (prepared by author).

step of a mid-scale plant construction, the above mentioned offtake agreements with airlines are important to convince the financial investors to provide further funding.

The actual timeline for SAF qualification is shown in Figure 8.2.

Additional projects for new SAF technologies are currently in the ASTM qualification process. Table 8.1 shows the status of completed qualification projects as of 1 August 2023. A successfully completed qualification project does not indicate the

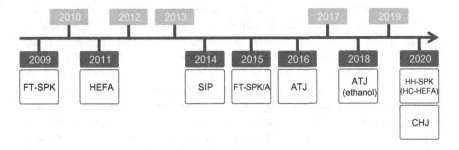

FIGURE 8.2   Timeline of ASTM SAF – certification.

Source: IATA, ASTM (prepared by author).

TABLE 8.1
ASTM D7566 Approved Conversion Processes

ASTM D7566 Approved Conversion Processes (as per 1 August 2023)				
Annex	Abbreviation	Conversion Process	Preferred Feedstocks	Maximum Blending Ratio
1	FT-SPK	Fischer–Tropsch hydroprocessed synthesised paraffinic kerosene	Coal, natural gas, biomass	50%
2	HEFA	Synthesised paraffinic kerosene from hydroprocessed esters and fatty acids	Bio-oils, animal fats, recycled oils	50%
3	SIP	Synthesised iso-paraffins from hydroprocessed fermented sugar	Biomass used for sugar production	10%
4	FT-SKA	Synthesised kerosene with aromatics derived by alkylation of lights aromatics from non-petroleum sources	Coal, natural gas, biomass	50%
5	ATJ-SPK	Alcohol-to-Jet synthetic paraffinic kerosene	Biomass from ethanol, isobutanol or isobutene	50%
6	CHJ	Catalytic hydrothermolysis jet fuel	Triglycerides such as oils from soybean, jatropha, camelina, carinata and tung	50%
7	HC-HEFA-SPK	Synthesised paraffinic kerosene from hydrocarbon hydroprocessed esters and fatty acids	Algae	10%

Source: ASTM, ICAO (2023) (prepared by author).

maturity of the technology. Rather, the qualification indicates that the SAF grade, when manufactured in accordance with the specification annex criteria, meets the requirements for safe flight operations and can therefore be used in the field.

The technological maturity of the new SAF process can be described according to the CAAFI Fuel Readiness Level (FRL). From ASTM certification to new plant construction, financing the project is a complex task that requires federal or state subsidies. With the US Inflation Reduction Act (2022), US companies have very quick access to subsidies that can be used to drive plant scale-up of new technologies. In the EU, the trilogue compromise on the introduction of mandates from 2025 is also leading to positive reactions from investors, as it is the first time that a minimum requirement has been formulated in Europe that creates a guaranteed demand for SAF. In the US, project development is focused on the two production processes HEFA and ATJ, as both technologies can be supplied with raw materials from domestic agriculture. In addition to new construction projects, this also includes the conversion of existing refineries to HEFA refineries, such as the Marathon refinery in Martinez, California, in a joint venture with NESTE. In Europe, oil companies are focusing on the conversion of existing oil refineries to HEFA refineries because this technology is close to conventional crude oil distillation and secures the existence of refinery sites (TOTAL, Shell, AGIP). In view of the strict organic feedstock requirements, projects for PtL production plants are currently under consideration in France, Germany and Sweden. While France can rely on nuclear power as an energy source, Sweden has sufficient hydropower for renewable electricity and Germany is relying on a massive expansion of wind farms and solar fields (Table 8.2).

In addition to the SAF specification ASTM D7566, ASTM has also approved the so-called "co-processing" under the JET A/JET A-1 specification ASTM D1655 with a blending ratio of 5%. This applies to HEFA feedstocks, FT feedstocks, and more recently, hydroprocessed biomass (at a 24% blend ratio). The co-processing takes place in the regular refinery process, with different feed points depending on the refinery technology used. Analogous to the ASTM D7566 raw material certification, the co-processing route also requires sustainability certification, as is the case with SAF for crediting in emissions trading. The advantage of the process is that existing refinery units can continue to be used for the production of middle distillates without additional capital expenditure. However, the EU blending quota of 6% in 2030 can only be achieved with the recently approved co-processing of hydroprocessed biomass. Co-processing with a slightly higher proportion of unprocessed biogenic feedstocks would also be technically feasible. Some refineries have already started to co-process biogenic feedstocks. A first certification as CORSIA eligible fuel was granted by ISCC-CORSIA in 2023. This means that co-processed biomass can also be counted as an emission-reducing fuel in CORSIA for international flights.

8.1 TECHNICAL SPECIFICATIONS FOR JET FUEL TO PROVIDE GLOBAL INDUSTRY STANDARDS

As a globally available and uniformly produced product, jet fuel requires uniform specifications from the outset. These specifications rely on the final jet fuel product properties to influence the crude oils used and their composition, the

TABLE 8.2
Fuel Readiness Level (FRL)

	Fuel Readiness Level (FRL)	
FRL	**Description**	**Toll Gate**
1	Basic principles observed and reported	Feedstock/process principles identified
2	Technology concept formulated	Feedstock complete process identified
3	Proof of concept	Lab scale fuel sample produced from realistic production feedstock. Energy balance analysis executed for initial environmental assessment. Basic fuel properties validated.
4	Preliminary technical evaluation	4.1 System performance and integration studies 4.2 Entry criteria/specification properties evaluated (MSDS/ D1655/MIL83133)
5	Process validation	Sequential scaling from laboratory to pilot plant
6	Full-scale technical evaluation	Fitness, fuel properties, rig and engine testing
7	Fuel approval	Fuel class/type listed in international fuel standards
8	Commercialisation validated	Business model validated for production airline/military purchase agreements. Facility specific GHG assessment conducted to internationally accepted independent standards.
9	Production capability established	Full scale plant operational
	Colour coding reference: Phase of development	light grey, technology phase; medium grey, qualification phase; dark grey, deployment phase

Source: CAAFI (2010) (prepared by author).

processing technology framework at the refinery, and the handling and storage of the jet fuel. These specifications ensure that jet fuel is safe to use regardless of where it is produced and the refining technology used. Jet fuel refining and processing is not standardised, which would require each batch to follow the same production process worldwide, but rather the final jet fuel properties are controlled to ensure consistent product characteristics. Each producer is free to choose the technology, but he can only market his product if it has the required characteristics and properties, for which there are ranges, minimum and maximum values. A JET A-1 is characterised not only by the product properties listed in the specification but also by several other parameters which are not relevant for the decision on conformity to the specification and can therefore have different values.

Furthermore, the JET A-1 specification follows a simple principle of the chemical industry: Only what is specified may be present, and what is not specified may not be present. This eliminates the possibility of other substances being contained in the jet fuel outside of the specification parameters that do not belong there and could potentially affect its storability and combustion in a jet engine's combustor. In daily practice, trace impurities are often found along the logistics chain from the refinery to the aircraft, which may constitute a threat to flight safety. Here are a few practical examples:

1. If JET A-1 is stored or transported in uncoated steel containers, rust particles are found in the jet fuel which, above a certain diameter, can potentially clog the fuel filters in the aircraft and thus impede the fuel flow to the engines. When rust particles bypass the fuel filters and enter the combustion chamber through the fuel nozzles, they can clog the fuel nozzles or have an abrasive effect on the combustion chamber and the engine blades, reducing engine life.
2. This also applies to any other debris that enters the high tanks through the ventilation system of older tank installations. Examples include plant seeds and windblown sand.
3. Inadequate drainage at the low point of elevated tanks will result in water accumulation as jet fuel releases water in storage. In the case of low point drainage, the free water can be drawn into the piping system of a tank farm and subsequently enter the aircraft. With water volumes of 200–300 L (50–80 USG), this can lead to engine operability issues or the formation of ice and associated fuel blockage to the engines.
4. And in addition, microbial growth (yeasts and fungi) can develop in the residual water in the aircraft tanks and clog the fuel filters on the engine. Residual water can also remain in piping systems from flushing operations and accumulate at low points. At high flow rates – when many aircraft are fuelling from a hydrant line at the same time – these water droplets can be entrained and enter the aircraft tank system.
5. Sand intrusion from the hydrant system is dangerous if the lines are not flushed carefully after construction work on the lines. Depending on the grain size, the sand may clog not only the fuel filters but also the emergency bypass system, resulting in spontaneous engine failure during flight.
6. If JET A-1 is transported in tanks that were previously filled with heating oil or that were cleaned from the inside with commercially available detergents and only superficially rinsed, the long-chain carbon molecules and/or detergent residues remaining in the tank compartment may impair either the high temperature or low temperature stability of JET A-1. The long-chain carbon molecules may not flow down to $-47°C$, as specified, but can freeze in the lines at temperatures as low as $-30°C$ and cause spontaneous engine failure. The detergents may affect the high temperature thermal stability of the fuel and form fuel system deposits that can impede fuel flow to the engine. It is therefore important to pay close attention to product quality and handling. The mere existence of regulations and procedures is not sufficient if the operating personnel do not know, do not understand, do not observe, misinterpret or deliberately ignore the regulations.

In the daily operational practice of the mineral oil industry, lessons have been learned from incidents over decades, and the conclusion drawn is that only close quality monitoring, careful selection and training of operating personnel, and consistent analysis of all causes of errors in the event of incidents can reduce the rate of "undesirable problems". With the increasing use of service providers and contractors who work with their own personnel on site or along the supply chain, this led to an extension of the monitoring and control requirements to these groups of people. These findings led to two main conclusions:

A. product quality and occupational safety are not conflicting issues, but two elements of a performance strategy, and
B. product quality and occupational safety are not competitive elements, but increase reliability and safety for all companies equally in networked production and supply chains throughout the industry.

Based on these principles, the oil companies formed joint committees to gather and share expertise and adopt common guidelines for the production and handling of fuels.

The first step was to agree on an international specification for jet fuel for the new generation of jet engines in the 1960s. The JET A-1 specification can be found in ASTM D1655 and in an equivalent product definition in the British military specification "Def Stan 91-091" (Defense Standard 91-091 of the United Kingdom Civil Aviation Authority). Based on these specifications, various national and military jet fuels were developed in different countries, each with its own designation, but with product characteristics that corresponded to the internationally common JET A-1. In Russia, the fuel TS-1 was similar to JET A-1, but could be used down to a freezing point of −60°C versus the −47°C of Jet A-1. In the US, a similar fuel, JET A, was used in civil aviation, which also corresponded to JET A-1 except for its reduced cold properties, with flowability down to −40°C. In military aviation, JP-4 (freezing point −72°C) and JP-8 (freezing point down to −47°C) are used. In addition, different grades of jet fuel differ in their respective flash points, i.e. the temperature at which the fuel will ignite when in contact with an ignition source (+38°C in the case of JET A-1).

The second step was to agree on voluntary quality and operational standards for the production, storage and transportation of JET A-1, as well as for the final implementation of aircraft refuelling. These standards are "living" documents that are constantly being updated to reflect new knowledge.

For the production, certification and transport of jet fuel, EI1530 is one of the industry's master documents. It contains "recommendations" for (1) aviation fuel quality assurance and traceability and agreement on uniform forms and documents, (2) sampling and testing of jet fuel, (3) requirements for certifying laboratories, (4) jet fuel production and production monitoring, (5) use of additives, (6) receipt, batching, certification and release of jet fuel batches, (7) storage design features and handling procedures for manufactured jet fuel, and (8) transportation features.

In order to develop such industry-wide documentation, the leading oil companies have agreed to delegate their expert engineers from chemical engineering, plant and vehicle construction, and occupational safety and environmental protection to industry-wide expert groups in which the experts exchange information, report on incidents, their causes and surrounding circumstances, and develop and evaluate options for action to prevent a recurrence of such incidents. This concerns the design of (1) facilities, (2) tanks, (3) pipelines and (4) means of transport, as well as (5) their daily operation and (6) requirements for redundant safety systems, emergency shutdown, alarms, etc.

In the third step, the experts from the oil companies looked at the application of the above fuel property and equipment design standards to the operational activities from the refinery to the aircraft. The group of experts developed the so-called "Joint Guidelines" under the Joint Inspection Group (JIG), which describe the necessary procedures for handling jet fuel and provide recommendations on how to proceed. These recommendations are not legally binding and can therefore be "interpreted" by individual oil companies and their facilities according to their structural or operational situation. Deviations from the recommended standard are therefore possible but not desirable. JIG 1 deals with operational practice guidelines for airport refuelling services, JIG 2 with operational standards for airport fuel terminals and hydrant systems, and JIG 3 with health, safety and environmental protection measures. These three technical publications now cover all aspects of a professionally managed and technically well-equipped jet A-1 supply chain from refinery to aircraft. For regional airports and commercial airfields with the so-called agency supply or through a fixed base operator (FBO), further JIG publications exist (JIG, 2021).

From an occupational health perspective, these publications address (1) personal protective equipment for employees, (2) the performance of physically demanding activities, (3) preventive accident prevention, (4) ergonomic design of stairs, ladders, safety equipment, etc., and (5) emergency response to incidents, accidents and environmental damage.

In the fourth step, the oil companies established a group of inspectors from the expert groups who inspect the airport fuel depots and the work of the fuelling service providers frequently every 6 months according to an internationally defined procedure. An inspection team consists of several engineers from the JIG member companies. It is strongly encouraged that the engineers alternate between inspecting their own facilities and those of their competitors. The inspection is carried out on the basis of questionnaires during normal operating hours. Not only the completeness of the documentation is checked, but also the execution practice is examined and the technical condition of the equipment is meticulously documented, down to checking the inspection dates on fire extinguishers, as an example. If an inspection reveals that the inspected company deviates from JIG procedures and recommendations, each deviation and each finding are recorded in writing. At the end of an inspection, the findings are discussed with the plant manager and his or her opinion is recorded. Depending on the urgency of each finding, either an unscheduled inspection is conducted to verify that the deficiencies identified have been corrected, or the findings of an inspection are recorded in the facility's inspection report. The follow-up check takes place during the next inspection, during which the chief engineer first examines the open findings from the previous inspection and then documents their correction.

A finding is carried over to the next inspection until it is fully resolved.

This procedure also allows for statistical evaluation of data from all inspection reports and provides indications of the need for modification and adaptation of technical or operational specifications. For a better understanding of emergency cases, here is a practical example: tank trucks and hydrant dispensers are equipped with various electrical safety devices to prevent accidents and fuel leaks. For example, an emergency stop circuit immediately shuts down the fuel pump if either the emergency stop switch on the vehicle is pressed or the aircraft refuelling operator releases the dead man's handle during refuelling. This was a switch that had to be held in the hand and kept constantly pressed. When the lever was released, the pump automatically shuts off. The refuelling operator is required to visually monitor the hydrant line and the filling line to the refuelling valve on the wing during refuelling. If fuel leaks from a hose or hose coupling, there is a high risk of fire because the aircraft's engines and wheel brakes are still hot after landing and will ignite the fuel on contact. Refuelling is done at a pressure of 4 psi, which is equivalent to a water column of 40 m (131 feet). During inspections, it was repeatedly found that aircraft refuelling operators did not hold the dead man's switch in their hand because it was annoying in the long run, but instead jammed the lever in the driver's door of the refuelling vehicle so that the lever was held in the safety position by the pressure of the door. Because the operators were not complying with the jamming restriction, the regulation was amended to require the operator to intermittently release and re-press the safety lever once per minute to ensure that this safety device is not disabled for convenience, thereby contributing to a fire hazard.

With the worldwide implementation of the Energy Institute and Joint Inspection Group safety standards, a remarkably high level of global safety has been achieved.

Regardless of the international airport at which an aircraft is refuelled, fuelling and aircraft refuelling are performed according to a uniform procedure with uniform safety standards.

As new companies enter the SAF production chain, the existing security architecture is challenged as the new entrants understandably focus on their SAF product rather than the established security practices of the oil industry. Companies in the start-up phase of the market face a tight liquidity situation during the market entry phase. It is not surprising, therefore, that spending on operational safety is limited to the mandatory minimum. In addition, in many cases, the workforce is recruited from other sectors and is overwhelmed by the high quality and performance requirements of jet fuel logistics. It is to the credit of the oil companies that they have achieved a globally unique safety standard through the continuous improvement of their quality management system, which corresponds excellently with the safety architecture in flight operations, without being required to do so by the authorities. The effectiveness of this double safety buffer is reflected in the global statistics on aviation incidents and accidents. Jet fuel is rarely found to be the cause of incidents and accidents, and in most cases, it is traffic accidents on the apron involving a tanker. This is also unacceptable, but in relation to the growth in global air traffic, fuel quality and aircraft refuelling as a cause remain consistently insignificant.

With the approval of the first synthetic jet fuel by ASTM in 2009, the civil aviation industry's concerns about reducing CO_2 emissions and the monopoly position of fossil jet fuel as the only fuel option for operating turbine-powered aircraft have increased. Initially, the approval had no impact on airlines' scheduled operations, as only sporadic individual flights were refuelled with the blend of JET A-1 and the newly approved FT-SPK (ASTM D7566, Annex A1) outside of the airport fuel depot infrastructure using separately filled tanker trucks. This changed with the second ASTM approval for HEFA–jet fuel. As part of its "burnFAIR" programme, Lufthansa fuelled a brand new A321 on the starboard side with a 50% JET A-1 HEFA blend (pre-blended by NESTE at its Porvoo refinery in Finland) and regular JET A-1 on the port side of the aircraft. From July 2011 until the end of December 2011, this aircraft operated four daily rotations with 8 flights between Hamburg and Frankfurt, totalling more than 1,100 flights. The aircraft was refuelled exclusively in Hamburg, with the HEFA blend stored in a rented tanker at the Hamburg seaport and transported daily to the airport by tanker. A second tanker was set up at the airport's fuel depot as an interim storage facility, from which the apron tankers then refuelled the starboard wing of the "burnFAIR" aircraft. During the 6 months of the research programme, the aircraft flew passengers between Hamburg and Frankfurt. Here, for the first time, the lack of an operating guide for handling synthetic fuel blending components became apparent as a need for action. As a precaution, the HEFA blend was not stored in the airport fuel depot. In addition, the HEFA blend had to be specifically filled into the wing tanks on the starboard side of the aircraft, which would not have been possible with commingled storage, as is common practice at airport fuel depots. The separate refuelling of the right wing tanks was carried out in accordance with all JIG regulations, which at that time only applied to the fossil fuel-powered JET A-1. In addition, the seaport fuel depot was declared to customs as a jet fuel depot under tax suspension, and all withdrawals were recorded in the accounting of the airport fuel depot. The seaport tank was also periodically sampled, with samples taken from the bottom of the tank, the centre of the tank and below the floating intake. As a result of the 6-month research programme, it was determined that a new operating procedure for handling SAF synthetic blending components would be required for the future supply of any airport fuel depot with an SBC. However, it's important to note that after blending of the SBC with the JET A-1 fuel, the resulting product is considered a JET A-1 fuel relative to all aspects of handling and operation. At that time, Lufthansa had already gained extensive experience in strategic self-supply with JET A-1, knew the relevant regulations for storage and handling of JET A-1, and was able to issue work instructions for the cockpit crews and the airport fuelling service provider based on the experience of its specialised operations engineers for jet fuel matters.

In the following years, the JIG and EI Committees increasingly dealt with the integration of SAF blends into the existing regulations, focusing primarily on the handling and transport of the SBC and the blending operation. It should also be noted that not every synthetic blending component is a sustainable aviation fuel. As mentioned earlier, a synthetic fuel can also be produced from coal and natural gas. Both feedstocks can be used to produce fuel, but the fuel produced is not sustainable as it does not reduce emissions according to US EPA RFS2 or EU-RED II.

The new regulations for the handling of SAF and SAF blends, developed jointly by the Energy Institute and the Joint Inspection Group, summarise for the first time the relevant regulations to be observed for the entire SAF supply chain from refinery to aircraft refuelling. Fourteen years after the approval of FT-SPK by ASTM, the aviation industry now has a consistent set of standards for handling SAF in the framework of the EI/JIG 1533 guidelines (Energy Institute, 2022).

8.2   QUALITY ASSURANCE REQUIREMENTS FOR SEMI-SYNTHETIC JET FUEL AND SYNTHETIC BLEND COMPONENTS (SBCs)

The new standards pay special attention to the determination of the fuel parameters of the blend, which must simultaneously meet the fossil jet fuel specification ASTM D1655 and the synthetic aviation turbine fuel standard ASTM D 7566, and which is stored as regular JET A-1 in an airport fuel depot, where it is subsequently blended with other JET A-1 quantities (Csonka et al., 2022).

Similarly, EI/JIG 1533 – Implementation Guidance considers the preparation of fuel blends and states that ASTM in its qualification procedures always assumes only one SAF component blended with one batch of JET A-1 (JET A-1 used in a blending process to be called "conventional blending component"). Such a binary blend is more in line with the arrangement in an experimental laboratory, but not with the reality of mass supply of jet fuel to airports. Due to the lack of or insufficient tank capacity at refineries and airports, there is an obvious tendency to simplify the blending processes of different SAF-SBC types with the conventional blending component JET A-1 as much as possible in order to be able to produce jet fuel blends efficiently with as few tanks as possible. In the opinion of the EI/JIG expert team, an uncontrolled release of all SAF/JET A-1 blending possibilities would contradict the basic principle of flight safety as a principle of action and lead to undefined conditions in airport fuel depots. The wording of the existing JIG regulations for the operation of airport fuel depots already stipulates that only jet fuel may pass the border of the airport fuel depot that is accompanied by appropriate documentation proving that the delivered quantity has already been tested as JET A-1 according to the specifications and thus fulfils the requirement for storage in the airport fuel depot. This requirement would not be met if SAF quantities were delivered to an airport fuel depot as SBC. SAF classified as SBC does not meet the JET A-1 specification and may not be accepted. Any blending of JET A-1 with SAF-SBC must therefore take place outside the boundaries of the airport fuel depot or in a dedicated and segregated area of the airport fuel depot in order to comply with the "Guidelines".

When SAF-SBC is blended with the conventional blending component JET A-1, the resulting blend is referred to as "semi-synthetic jet fuel" or synthetic aviation turbine fuel (SATF).

It should then be noted that such a "semi-synthetic jet fuel", which can be declared as "JET A-1" if the parameters for neat JET A-1 are met, in turn serves as "conventional blending component JET A-1" in a subsequent blending process with another SAF-SBC, although it is no longer neat JET A-1.

In this constellation, a "SAF-SBC is mixed with a SAF-SBC/JET A-1". In this case, a distinction would have to be made between two identical SAF-SBCs or two different SAF-SBCs, which in principle makes no difference to the procedure as long as the final blend remains within the JET A-1 specification and the total SAF blends do not exceed the maximum blending ratio of 50%. A two-step blending operation, regardless of the order of the blending components, can in principle be described as a "multi-blend".

The following example calculation illustrates the procedure and its hidden limitations:

A batch of jet fuel is to be blended with 30% HEFA. Due to the differences in density between HEFA and JET A-1, the following calculation is expressed in the weight unit "metric ton" for a better understanding of the product relationships:

1. There are 5,020 tons of JET A-1 in the blending tank. 20 tons is the remaining quantity not reached by the floating suction in the tank space. The possible discharge is 5,000 tons.
2. To obtain a 30% blend for the batch, 2,143 tons of HEFA are added. The tank then contains 7,143 tons of semi-synthetic jet fuel in the tank plus the 20 tons remaining at the bottom of the tank (which is not included in the following calculation).
3. The 7,143 tons can be divided into 70% fossil JET A-1 (5,000 tons) and 30% HEFA (2,143 tons). This situation reflects a typical binary mixture.
4. In a further step, another 3,000 tons of ATJ-SBC will be added to 7,143 tons already in the tank. These successive additions of SAF components will be referred to hereafter as "stacked blending" because the product is expanded by one component at a time.
5. The new total is 10,143 tons of semi-synthetic jet fuel. In terms of the second blending stage, the ATJ blend is approximately 29.6% of the 10,143 tons. However, since the ATJ-SBC was blended on top of an already pre-blended amount of semi-synthetic jet fuel, the combined amount of SAF-SBC from HEFA and ATJ is 5,143 tons (2,143 + 3,000 tons), or 50.7%. This total blend could theoretically still be within the limits of the ASTM D 1655 specification, but the blend is above the 50% maximum and is therefore not permitted.

This example illustrates the problem of an uncoordinated blending process outside an airport fuel depot.

Precise compliance with the blending rules is a safety measure that must be observed by all participants in the SAF supply chain:

1. For "stacked blends", each individual SAF-SBC may only be blended up to its own maximum blending limit. This rule was followed in the example.
2. If a conventional blending component does not consist of neat JET A-1, but is already a semi-synthetic jet fuel because an SAF-SBC has already been blended into it upstream, this fact must be known prior to the beginning of the subsequent blending process. This rule was not observed in the example.
3. The introduction of a different SAF-SBC in the second blending stage than the one used in the first blending stage is not permitted in a certification according to Def Stan 91-091 because the components are only specified in percentages and not in absolute quantities and products. In this case, only another blend with HEFA should have been introduced.

However, it would have been possible to produce two binary blends, one with HEFA as SBC and neat JET A-1 as "conventional blend component" and the other as ATJ-SBC also with another neat JET A-1 quantity. Thus, each binary blend should contain a maximum of 50% of the SBC and would have to be within the ASTM D 1655 specification on its own. Under these conditions, both blends could have been delivered to an airport fuel depot and stored in a commingled storage tank.

When exporting SAF-SBCs or semi-synthetic jet fuel (SAF-JET A-1 blends), it is important to know what documentation and data is required in the recipient country. This is especially true for exports to countries where Def Stan 91-091 is required by law, as this certification requires additional measurements that are not collected according to ASTM D 1655 and Canadian CGSB 3.23. In the worst case, the import would be stopped at the border due to incomplete documentation and missing test certificates. During the production of SBCs, additives should be added to the SAF-SBC as soon as possible to prevent peroxidation and gumming. In this respect, SAF producers must add additives to their products as soon as possible and as well as possible. The final location of the SAF-SBC must be determined in advance, as the additives differ depending on the jurisdiction of the applicable specification. For example, Def Stan 91-09 requires the addition of static dissipative additives, while ASTM D 1655 and CGSB 3.23 do not. In practice, the impression is always given that SAF does not contain any additives and is therefore dependent on sufficient total additive addition of the JET A-1 component, or that the fossil JET A-1 quantity must be over-added on the refinery side to achieve sufficient additive addition of the blend. EI/JIG 1533 clearly counters this impression and thus improves the starting point for future direct sales of SAF to airlines, once the use of neat SAF without a blending requirement has been decided by the international bodies.

If semi-synthetic jet fuel is to be produced in an airport fuel depot, EI/JIG 1533 requires that this area be completely separated from the normal operations of the depot. Only after an SAF blend has been blended in accordance with regulations, and the blend has been pre-treated to achieve the homogeneity of an in-line blend at a refinery site, and all laboratory tests confirm that the blend meets ASTM D 1655, can the blended quantity be pumped from the segregated area into the storage area of the airport fuel depot.

With these specifications, the EI/JIG working group provides the first set of guidelines for handling SAF and eliminates the existing uncertainty as to how an SAF-SBC can be safely placed on the market in accordance with the technical parameters. This first edition of EI/JIG 1533, dated 22 November 2022, will certainly be supplemented and detailed in the coming years. This will include how CHJ, which is currently the only SAF grade that already meets the requirements of ASTM D1655, but is also currently only approved up to a blending level of 50%, can then be placed on the market. In addition, the 100% neat SAF option promoted by airframe manufacturers will also need to be addressed in this directive. It remains to be seen whether Boeing's intention to blend synthetic diesel with fossil JET A-1 will be realised and which regulations will have to be observed in this case. The spectrum of SAF blends remains complex and confusing. It is therefore all the more important that the new market participants in the SAF production chain also behave in a system-compliant manner and contribute to the further improvement of the safety standard.

8.3   REGIONAL DIFFERENCES IN FEEDSTOCK AVAILABILITY

With respect to Section 8.7 and feedstock supply, another factor must be considered when looking at SAF production: With the exception of domestic air transport in the US, which is located in the proximity of the agricultural area for energy crops and therefore takes place within the same jurisdiction and allows for manageable transport distances between feedstock production and the SAF production site, the global demand for SAF is located in regions where large-scale production of biogenic feedstocks is not possible (e.g. Singapore) or is politically undesirable (e.g. Germany). This means that the production areas for biogenic feedstocks, the SAF production sites and the busy airports, which will be the transshipment points for large quantities of SAF in the future, are located at a great distance from each other, which has a negative impact on production costs, the functioning of the supply chain and the involvement of the responsible governments. On the one hand, the EU27 is introducing mandatory blending quotas from 2025, which will increase over time, while on the other hand, many governments assume that the required SAF volumes will be largely covered by imports, without knowing where these imports will come from. By relying on imports, governments are relieving themselves of their core responsibility of providing their own citizens with the "goods of daily use". This attitude illustrates a core problem of energy supply: There is no consistent planning based on a working hypothesis of the likely evolution of demand. Only on the basis of this analysis can the demand for SAF production capacity be determined in the run-up to the market, with each government sensibly assessing this capacity for its own national territory. This statistic remains independent of the question of whether the capacity is to be provided domestically or built in third countries. The technology mix must then be determined as a result of the availability of raw materials. When such surveys are conducted by supranational organisations such as ICAO, it would be advisable to share the results with professional project developers and investors at international SAF conferences, of which there are now many, in order to accelerate the process of production expansion. Looking back over the past 14 years since the approval of the FT-SPK, the scope of regulation of what is undesirable as a raw material or product in the EU has increased significantly, without any indication of what would need to be done as a positive solution to the problem in order to further the goal of GHG reduction in aviation.

The magnitude of the problem, the increasing pressure to act and global warming demand that all parties contribute to solutions rather than obstruct them.

With the Inflation Reduction Act and accompanying legislation, the US government has injected new momentum into the SAF issue, reviving the proverbial "frontier spirit" of the US. The success of the globally standardised JET A-1 specification paved the way for global aviation in the jet age. The challenge is to repeat that success 60 years later.

8.4 THE FUTURE ROLE OF RFNBOs (RENEWABLE FUELS OF NON-BIOLOGICAL ORIGIN)

The future role of RFNBOs will depend on three main factors:

1. Green power must be available in sufficient quantities to enable uninterrupted 24/7 production.
2. The price of green electricity must be set as close as possible to the cost of production to make the end product, PtL jet fuel, marketable.
3. Nuclear power must be included as a source of electricity.

The industrial decarbonisation strategy is largely based on green electricity in direct use or as a raw material source for the production of green hydrogen. Both products are highly demanded in a competitive market, as many manufacturing companies want to switch their production to green electricity or green hydrogen as an energy carrier. In addition, the share of electric vehicles in Europe and the US is increasing, as is the use of electric heat pumps to heat buildings. It is still not clear which technologies will be able to produce this amount of electricity, whether these plants will be available in sufficient quantities, and whether the grid infrastructure will be sufficient to overcome the transport distances between electricity generation and electricity consumption. In any case, there will be competing uses, which are usually resolved by market price.

For airlines, the next step will be the substitution of the fossil JET A-1 with SAF, followed by SAF RFNBOs, followed by electric aircraft engines powered by electricity. By 2050, the majority of the SAF supply will come from biomass and waste. Electric aircraft engines will be used only marginally until 2050 due to the lack of technological maturity, and then initially only for short-haul flights. Hydrogen as a substitute fuel faces the same problem as electricity: Competition for use in the event of excess demand is also determined by price. In the period up to 2050, for development reasons, only short-haul commuter aircraft will fly with pure hydrogen propulsion. The main source of propulsion in this option remains the combustion engine with hydrogen as the energy carrier, which is converted into PtL jet fuel and such fuel used for aircraft refuelling.

The public expectation for PtL jet fuel is analogous to the technology maturation and manufacturing cost development that has occurred in other industries, particularly consumer products. The simplistic message is that competitive pressure makes products more efficient, and at the same time, the price falls as a result of competitive pressure or diverging consumer preferences. This approach falls short in the case of RFNBOs because they are used as a means of production and not as an end product. Jet fuel in general and PtL jet fuel in particular are not perceived by passengers as a means of production. While every passenger is aware that an aircraft cannot fly without fuel, the magnitude of the conversion from fossil jet A-1 to eSAF is not tangible.

From a production point of view, it places technological development at the limits of the laws of nature, which literally represent a natural limit of development beyond which there is no further improvement. The assumption that analogous efficiency gains will take place in battery propulsion and in the conversion of water into hydrogen is therefore optimistic in view of the natural limits of development and distracts from the core problem: The resource consumption for wind turbines and solar panels, for pipeline networks and electrolysis plants must be included in the life cycle analysis from an objective point of view. Then there are the costs of hydrogen storage, which are significantly higher than the simple storage of synthetic fuels in above-ground tanks.

8.5 FURTHER BLENDING REQUIREMENTS AND RECOMMENDATIONS

With regard to the blending of jet fuel and SBCs, an extensive research and investigation programme was launched under the project name "DEMO-SPK" on behalf of the German Federal Ministry of Transport and Digital Infrastructure from 2016 to 2018, in order to test all conceivable blending sequences with different SAF-SBCs in practical tests and to document deviations from the expected results as comprehensively as possible and to determine their causes (Bullerdiek, 2019). For this purpose, 600 tons of a multi-blend consisting of fossil JET A-1, HEFA-SBC and ATJ-SBC were produced, and the possible blending processes were tested in extensive laboratory tests.

The findings and recommendations derived from the DEMO-SPK project for future blending operations can be summarised as follows:

Since individual batches of conventional jet fuel from different sources are usually transported separately, but upon arrival at the airport fuel depot are stored together within the airport supply infrastructure, the physical separation of the delivered batches no longer exists. This inevitably leads to mixing and blending. In most cases, this is caused by pumping operations at the airport fuel depot, as special mixers are not usually installed in the receiving tank. Since only specification-compliant JET A-1 may be used, this is formally permissible. However, due to the mixing behaviour and compatibility of JET A-1 blends (binary and stacked blends, if applicable) containing varying proportions of different types of SPK, "multi-blending" should preferably be performed in tank farms equipped with appropriate mixing equipment.

1. Multi-blend JET A-1 is a blend of the conventional blending component JET A-1, which conforms to ASTM D1655, and at least two other ("multi") SBCs, which conform to ASTM D7566. The SBCs may be from the same ASTM D7566 Annex or from different ASTM D7566 Annexes. It can be assumed that "sequential blending" (EI/JIG 1533) is the most common practice outside of refineries. In all cases of sequential blending, the main concern is the homogeneity of the blend. The blending tank must be equipped with the necessary technical equipment, i.e. tank-side entry mixers and/or tank recirculation mixing systems.
2. In fact, depending on the parameters and characteristics of the conventional jet fuel used as the blending component base, the maximum effective blending ratio may be lower than the approved ASTM certification limits.

Such restrictions to a lower than maximum allowable blending limit due to composition and fuel property parameters should be observed, recorded and communicated in the appropriate documents and airport fuel depot IT systems.

3. Regarding the variety of sustainable aviation fuel types (FT-SPK, HEFA-SPK, SIP, SPK/A, ATJ-SPK and others) already qualified by ASTM, EI/JIG 1533 provides the first set of advice for handling multiple blended SPKs. The safest way to avoid any off-spec fuel situation is to start with binary blends of JET A-1 as the CBC and start with an initial SAF-SBC to create such a binary blend called "semi-synthetic jet fuel" (EI/JIG 1533). Further blending of a semi-synthetic jet fuel into a "multi-blend" jet fuel in a subsequent secondary process is based on the fuel properties of the now existing semi-synthetic jet fuel acting as the "conventional blend component". In the case of a stacked blending procedure with a conventional jet fuel component JET A-1 and two or more SAF-SBCs, EI/JIG 1533 remains cautious as further practical experience needs to be shared. In the DEMO-SPK project 2016–2018, this was one of the research objectives. The "stacked" blending approach adopted in the DEMO-SPK project required extensive testing of the semi-synthetic jet fuel after the required homogeneity and settling time had been achieved. Only after successful completion of all blending and testing procedures, this semi-synthetic jet fuel batch was subjected to another blending procedure with another SAF-SBC. Of course, the blending ratio and the total SAF-SBC content in the blend were documented. As long as the multi-blend semi-synthetic jet fuel was within the ASTM D1655 specification and the maximum allowable blends per ASTM D7566 were adhered to individually and collectively, no difference could be detected in the final comparison to the blend of two previously prepared binary blends with the same blend ratios. However, from a scientific point of view, this is not yet conclusive proof that stacked blending, performed and documented in a controlled manner, has exactly the same properties as the combination of binary blends to form a multi-blend. However, the research results indicate that stacked blending, even with different SAF-SBCs, basically simulates a mixed storage situation in an airport fuel depot, where different SAF-SBCs in different blending ratios and different CBC JET A-1 from different production processes are stored together. The advantage of stacked blending is the unchanged composition of the JET A-1 batch, which is likely to be the exception when binary blends converge.

4. EI/JIP1533 does not permit two SAF-SBCs to be blended together to form a binary SAF/SAF multi-SBC belonging to two different ASTM D7566 annexes and then blended with JET A-1 as a conventional blending component!

However, the procedures outlined in Subsections 8.2 and 8.3 require a number of storage tanks and are likely to be both costly and time consuming. In a more likely scenario, a blender blends such SPKs in a storage unit as "stapled blends" and determines the fuel properties at each blending stage. Based on such figures, the maximum blending ratio with dedicated batch of JET A-1 can be calculated. It may be good practice to deduct a safety margin from such determined maximum blending ratios in order to avoid any risk from a multi-blend batch of jet fuel that does not meet the requirements of the ASTM D7566 and ASTM D1655 specifications and may cause a safety issue when used for aircraft refuelling purposes.

5. For practical reasons, consideration should be given to starting the production of semi-synthetic jet fuels only in fixed blends. Such fixed blends may be well below the observed maximum blends and will not relieve any operator of the need to test fuel properties accurately, but will provide a global database on the behaviour of fuel properties as a function of blend ratio to encourage the further development and use of sustainable aviation fuels. In addition, such given blends may facilitate the handling of SAFs in general as well as tank farm operations due to the revolving parameters. Multi-blend semi-synthetic jet fuels shall be approved for use provided that the JET A-1 composition has been verified and recorded prior to blending and that the SPK components have subsequently been blended in compliance with the final multi-blend specification requirements to be achieved (i.e. max. 70 % JET A-1 topped with 15 % HEFA + 10 % ATJ + 5 % SIP). With reference to paragraph (3) above, the EI/JIG shall determine how the multi-blend semi-synthetic jet fuel is to be produced.

6. The Declaration of Conformity shall be amended to state that the characteristics of the fuel batch meet the specifications for jet fuel as set forth in ASTM D1655 and that each SPK type used has met the specifications of ASTM D7566 prior to the commencement of blending.

7. Multi-blend semi-synthetic jet fuel shall not be delivered to an airport fuel depot unless it is fully certified prior to reaching the airport or depot boundary. The multi-blend semi-synthetic jet fuel shall be delivered with a Certificate of Origin issued by the blending facility listing all SPKs and their percentages in the final blend as well as the percentage of fossil jet fuel in such blend. The format of such a "Blend Certificate of Origin" has yet to be developed as a template.

The exemption described in EI/JIG 1533 for blending SAF and JET A-1 into a semi-synthetic jet fuel in a segregated area within an airport fuel depot should be strictly regulated for safety reasons. The operator of an airport fuel depot must in no way be the (technical) blender of jet fuel in the segregated area of an airport fuel depot (i.e. the actual blending process must not take place in an airport fuel depot). In accordance with the four-eye principle, the responsible manager producing a semi-synthetic jet fuel in a segregated area of the airport fuel depot should not be the receiving operator for deliveries to the airport fuel depot. In addition, the conventional blending component JET A-1 should not be withdrawn from the airport fuel depot stock, but should be delivered directly to the separate blending facility as neat JET A-1. It must be possible to avoid using a residual quantity of another semi-synthetic jet fuel already in storage as the basis for the blending process. For this purpose, independent storage lines must be available for filling the blending tanks in the separate area of the airport blending depot.

Currently, new synthetic fuel types are being qualified by ASTM that contain aromatics (Kaltschmitt and Neuling, 2018). Based on these conditions, additives must be added to the SPK and the final SAF composition will meet the ASTM D1655 specification without further blending. As they meet the specification, such SAF can be declared as JET A-1 (according to the defined freezing point or even better). These pure synthetic jet fuels will henceforth be referred to as "Ready-for-Use SKA". For safety reasons, ASTM may also require a blend of such "Ready-for-Use SKA" until their reliability in daily airline operations is demonstrated.

Matching the specification does not necessarily mean that all parameters of the fuel types are identical, but that is why the ASTM D4054 qualification process requires evaluation of all of the jet fuel parameters prior to achieving qualification.

8. A fully synthetic jet fuel JET A-1 shall not be used as the basis for further blending with an SAF-SBC unless explicitly approved by the applicable standards. JET A-1 used as the basis for blending shall be a fossil jet fuel conforming to ASTM D1655. With regard to the Ready-for-Use SPK, it needs to be clarified whether this provision (ASTM D1655) also applies when the base fuel for blending is a pure synthetic jet fuel. According to ASTM D1655 and ASTM D7566, ready-for-use SKAs meeting both specifications can be blended without restriction. A fuel tank may contain shipments of semi-synthetic jet fuel (SPK SBC blends with CBC JET-A1, as well as various ready-for-use SKAs and neat JET A-1). If additional operational handling requirements are to be applied, the JIG guidelines will be amended accordingly. It is recommended that if a depot operator detects any deviation from expected behaviour or observes any out-of-spec situation associated with the storage of fully synthetic JET-A1, he should request immediate notification.

9. The JIG Guidelines shall be amended accordingly to include a provision that Airport Fuel Depot Operators shall track the receipt, storage and distribution of SPK semi-synthetic jet fuel, fully synthetic jet fuel and neat JET A-1 in their inventory management system.

10. The airport fuel depot operator shall monitor and track maximum blending rates in accordance with ASTM D7566.

11. In addition to the existing volumetric monitoring of jet fuel quantities, the inventory management of an airport fuel depot shall include a second level reflecting the ownership and quantity of the portion of SAF delivered within the total batches recorded. The use of SAF shall be based on documented agreements between shareholders or throughputters and their customers that deal with a percentage of sustainable aviation fuel in each fuelling operation for that customer. Alternatively, a preferential allocation of such virtual inventory to aircraft fuelling may be made in such a way that such virtual inventory is delivered first to the defined customer until its ordered

quantity of Sustainable Aviation Fuel has been consumed. Since such fuel-ling data movements relate only to virtual consumption, there is no need to track maximum blending rates as specified in ASTM D7566.

8.6 BOOK & CLAIM VERSUS TRACK & TRACE

In order to be able to use SAF-SBCs and SAF as a fully synthetic jet fuel in the EU-ETS or globally in international aviation (CORSIA) adequately and seamlessly in the future, it is necessary to establish suitable concepts for accounting for SAF in GHG regulation systems such as the EU-ETS or CORSIA. If an airline chooses to reduce its emissions by using SAF, the two options (a) "Track & Trace" and (b) "Book & Claim" allow for effective accounting of such fuels in a GHG monitoring and regulation system.

8.6.1 TRACK & TRACE (MASS BALANCE ACCOUNTING)

Based on a fuel purchase contract for SAF (either as SAF-SBC in a semi-synthetic jet fuel or by purchasing Ready-to-Use SKA), the physical storage and transportation of this jet fuel must be monitored from its point of production to the point where the product is fuelled into an aircraft. At that point, it is considered to have been con-sumed. Proof of such fuel delivery must be provided to both the aircraft operator and the seller of SAF as SAF-SBC or SAF fully synthetic jet fuel. The seller of the SAF-SBC can then issue a certain number of CO_2 certificates, which allow for a reduction in the amount of CO_2 saved by using sustainable instead of conventional jet fuel. As part of the SAF purchase agreement, the purchasing airline – as the buyer of such jet fuel – is entitled to receive the CO_2 certificates issued by the seller. The airline can then decide to use these CO_2 offsets to reduce its CO_2 offset requirements, thereby avoiding the need to purchase the same amount of carbon credits. Alternatively, the airline may sell the carbon credits to other parties that wish to purchase them.

8.6.2 BOOK & CLAIM

The Book & Claim approach requires the separation of the physical fuel and its sustainability characteristics, which are transferred to tradable certificates (called Guarantees of Origin) at a certain point in the supply chain. For a certain amount of SAF semi-synthetic jet fuel or SAF-SBC that is injected into the fuel supply sys-tem, the buyer of SAF volumes receives a value equivalent number of Guarantees of Origin (the number of Guarantees of Origin must be equal to the amount of SAF produced, sold to a customer or sold to an airline). Based on a valid and binding purchase agreement, these Guarantees of Origin can be transferred to the airline as the buyer, as long as the supply chain is secured against tampering and the move-ment of quantities takes place in a closed loop. The airline will use the Guarantees of Origins to account for SAF purchases and associated emission reductions in a GHG regulatory system. By redeeming a purchased Guarantee of Origin, an air-craft operator declares the amount of jet fuel purchased and received to count for SAF, regardless of its actual origin. The responsible airline or fuel distributor shall demonstrate to the competent authority that an equivalent quantity of SAF, meeting the relevant sustainability criteria, has been supplied to the fuel supply system and thus burned in an aircraft under the applicable volume movement and monitoring conditions of the EMCS. The airline, as the end user, pays a higher market price for SAF than for fossil A-1 jet fuel and receives a number of Guarantees of Origin in proportion to the SAF purchased to compensate for the price premium. Theoretically, the SAF can be sold at a slightly higher price than the fossil jet fuel, with the producer/supplier covering its higher production costs by selling the Guarantees of Origin separately. However, the market for CO_2 certificates must match the price of the guarantee of origin. There is a time lag and uncertainty in this alternative, as the price of CO_2 certificates can fluctuate in a volatile trading period (Bullerdiek et al., 2019).

8.6.3 Implications with Track & Trace (Mass Balance Accounting)

Under the ICAO-agreed tax exemption scheme, tax-free jet fuel must be segregated from all other taxable fuels and such segregation, storage and transportation must be accounted for by the producer and all participants in the supply chain until the fuel reaches the airport fuel depot. All deliveries of jet fuel will be commingled, and the aircraft fuelling companies will take the jet fuel from the depot and deliver the tax-free product to the airlines' aircraft for refuelling. Quantities delivered to aircraft are reported daily to the airport fuel depot operator. Quantities leaving the depot must match the quantities delivered to aircraft with a daily zero balance. Under these conditions, jet fuel temporarily stored in bowers or in the hydrant system is still considered part of the depot's inventory, despite the fact that the quantities have physically left the storage tanks (known as "dead stock" in the case of a permanently filled hydrant system). Following the principles of tax-exempt jet fuel and its supervision by the relevant fiscal authority (customs), the competent authority could also trace the supply chain of sustainable aviation fuel from its production in a refinery or blending facility to its consumption in an aircraft and verify the number of CO_2 certificates as well as the amount of CO_2 savings represented by these certificates by subtracting the volume of jet fuel consumed from the volume of jet fuel produced. The remaining amount of fuel would correspond to the data in the inventory management system of the airport fuel depot operator and its affiliated aircraft refuelling companies, which also maintain inventory in their fuel trucks. As with tax-exempt conventional jet fuel, all movements, intermediate and final storage of sustainable aviation fuel at an airport fuel depot must be continuously monitored and controlled prior to consumption. While this monitoring can be easily done in the European Union with the Excise Movement and Control System (EMCS), a supranational, standardised electronic system for excise tax monitoring, i.e. monitoring national and international movements of aviation fuel in EU member states, it may be quite different in other regions of the world without such a monitoring and control system. However, if an airline intends to count its SAF consumption towards its obligation to purchase CO_2 certificates, it will most likely request certificates for those airports where the airline has entered into SAF purchase agreements and where SAF has been used to refuel aircraft accordingly.

In addition to the aspect of different regulatory frameworks and legal systems with their individual provisions for renewable feedstocks, accounting methods and requirements, the method of issuing certificates must be accepted by the tax authority of the airline's state of registry. In the absence of a monitoring system covering fuel movements, a reliable monitoring procedure will be mandatory to track and control the production of SAF from well to wing.

Authorities consider the airport fuel depot as the central collection point for information regarding the receipt of SAF and its distribution to dedicated airlines based on their SAF purchase agreements. The Airport Fuel Depot will collect all transport and intermediate storage related information from production/blending to storage in the depot's tanks as the "downstream". This also applies to the allocation of SAF quantities to selected aircraft refuelling operations performed by aircraft refuelling companies and their meter readings of jet fuel supplied to such airlines ("wingstream").

Assuming that the physical molecules of SAF cannot be traced in either a blended storage or a wingstream delivery from the depot tanks to the fuel inlet valve in an aircraft wing, it becomes essential that the allocation of SAF quantities to an airline be done on a bookkeeping basis at the depot. In practice, this should be done electronically in the depot's fuel inventory management system, which is already approved and frequently inspected by the relevant tax authority (customs) for tax exemption reasons, but also to record, document and verify individual sustainability and emission reduction properties during inspections.

In order to provide the necessary transparency, a second inventory level of the fuel inventory management system can split the stored volume of each shareholder/throughputter into conventional (fossil) jet fuel (as before) and SAF. Based on the specification of the SPKs, the SAF volume has to be split into different SPKs if these SPKs have different CO_2 accounting values/emission factors (even though the feedstock types may be identical)! While the fuel receipt and consumption in the first level is based on meter readings, the receipt and consumption in the second level can only be defined by delivery documents (e.g. a certificate of origin as proposed) and by the provisions of the fuel purchase agreement/refuelling agreement between the seller (shareholder/throughput operator) and the buyer (airline)!

The consolidated fuel movements between the first and second inventory levels of the fuel inventory management system must be identical. Based on the physical movements (Downstream and Wingstream, first level), the movements of the second inventory level reflect the same magnitude of volume movements, but only related to the sustainable and conventional jet fuel quantities on a virtual consumption pattern, i.e. preferred consumption of SAF as agreed in the supply agreements between airlines and suppliers.

8.6.4 Implications of Book & Claim

The implementation of an EU-wide Book & Claim concept and the development of instructions for its day-to-day operation will require interaction between the European Commission and other institutions.

The implementation of a Book & Claim concept on a global scale is conceivable and reasonable. It requires provisions on how to recognise different standards (e.g. EU-RED vs. US EPA) outside their primary jurisdiction. Otherwise, aircraft operators will only be able to uplift SAF that meets the accounting rules of their home country (State of Registry), which means that they are likely to only uplift SAF at domestic airfields/hubs and any uplift offshore would not qualify for the issuance of (domestic) Book & Claim certificates unless expressly ruled in a global Book & Claim system.

The issuance of a Guarantee of Origin requires proof of sustainability for the raw materials used to produce a particular batch of SAF. The appropriate sustainability documentation must be available throughout the supply chain up to the final production of the SAF. In addition, in order to ensure a closed chain of custody from feedstock production to the final use of SAF in an aircraft, it is necessary to ensure that the supply chain is certified according to recognised sustainability certification schemes up to the point where a Guarantee of Origin is issued based on a Proof of Sustainability.

A Guarantee of Origin must be issued as long as the pure synthetic content of a binary blend or multi-blend semi-synthetic jet fuel is known or can be derived from appropriate documentation such as a Refinery Certificate of Quality. As SAF is transported along the supply chain, it is mixed with other batches of fuel. The complexity increases with the number of times a particular batch of semi-synthetic jet fuel has been blended, as it may be difficult to determine the synthetic content based on (transport-related) fuel documents. Guarantees of origin should therefore be issued as soon as possible after blending.

A Book & Claim approach must prevent unintentional multiple counting of SAF, i.e. the improper issuance of Guarantees of Origin without a corresponding sustainable fuel being produced and placed on the market for use in an aircraft. Book & Claim concepts are already embedded in other sectors, such as the energy business, which deals with the marketing of renewable energy and biomethane. When implementing a Book & Claim concept for SAF, the experience in other sectors can enhance the functionality and manageability of such a system and facilitate its implementation in the aviation sector.

8.7 SAF LOGISTICS

SAF Producers must define within their business model which role(s) they want to play in the marketing of their SAF product:

B2B: The SAF producer sells its product as an intermediate product to oil companies that produce JET A-1. In these cases, the SAF producer has no sales contact with the airlines as end users. The SAF producer's product is blended at the buyer's facilities and delivered to the airport fuel depots as a semi-synthetic jet fuel known as JET A-1. In this scenario, the SAF producer's product liability is only towards his buyer, to whom the SAF becomes the property and for which the buyer is subsequently liable.

B2C: The SAF manufacturer sells his product to the end user, i.e. the airlines. The SAF producer has an appropriate sales organisation which ensures that the producer obtains and maintains a permit from the airport authority to market jet fuel at the airport. This also applies if the SAF producer distributes an SAF-SBC which is blended with neat JET A-1 in a separate part of the airport fuel depot and subsequently stored as semi-synthetic JET A-1 grade jet fuel in the tank farm. Prior to blending, the SAF producer is liable for any damage attributable to its product. After blending, the seller of the semi-synthetic jet fuel will be held liable for any incident caused by or attributable to its blended product, as its conventional jet fuel component will continue to be the predominant product in the blend.

If the SAF producer has an ASTM approval for its SAF grade, which allows the product to be stored directly as neat SAF and fuelled into aircraft without blending in the sense of a "ready-for-use SKA", it will also need a distribution license from the airport authority and a contractual agreement with the airport fuel depot for throughput or participation.

Due to the necessity to comply with the specification limits according to ASTM D1655, SAF-SBC must be blended with JET A-1 prior to aircraft refuelling, whereby – as explained – the maximum achievable blending ratio is determined by the parameters of the respective JET A-1 batch and the blending process must therefore be subject to permanent expert monitoring for quality assurance reasons. This requires that the physical transport paths of SAF-SBC and JET A-1 are brought together in an intermediate storage facility prior to final removal for aircraft refuelling. Otherwise, it is not possible to prove the quality of the blend prior to the final withdrawal. The following options exist for minimizing the "downstream" transportation costs of the two products:

Option 1 (Refinery Blending):

Transport SAF-SBC from the biorefinery to the oil refinery with blending at the oil refinery. Subsequent use of the above-mentioned downstream logistics for blending.

Advantages: (1) Quality assurance by the refinery operator and (2) possibly more favourable transportation costs due to unit cost degression in mass transportation in already existing transportation chains.

Disadvantages: (1) The agreement of the refinery operator and the oil companies involved is required. (2) Monopoly pricing by the refinery operator for blending. (3) Change of ownership for the JET A-1 blend already "ex refinery" and thus shift of capital commitment for the product from the oil company to the airline with corresponding financing costs until it is used in the aircraft. (4) In addition, the physical transport chain for the SAF-SBC is extended with a corresponding deterioration in the life cycle assessment.

Option 2 (Neutral Blending Depot):

Transport the SAF-SBC and the JET A-1 from the respective production sites to a common intermediate storage facility between the production sites and one or more airports.

Advantages: (1) Independence from oil company approval requirements. (2) Achieve competitive blending prices. (3) Possible reduction in transportation distance for SAF-SBC with improved transportation costs and (4) associated improvement in life cycle assessment compared to the refinery blending option.

Disadvantages: (1) Additional storage and handling for SAF-SBC and JET A-1. (2) Transfer of ownership for JET A-1 already in intermediate storage "into tank" with corresponding return of capital of the inventory by the airline and (3) possibly longer transport routes with associated additional costs.

Option 3 (Airport Blending):

Separate transportation of SAF-SBC and JET A-1 to the airport with physical segregation of a portion of the airport tank farm as a "blending pre-storage" and subsequent transfer of the blend to the airport tank farm.

Advantages: (1) Avoidance of additional costs for JET A-1 transport and direct SAF-SBC transport to the airport. (2) Synergy potential in the operation of the blending tank farm by the airport tank farm operator. (3) No need for approval by the oil company selling JET A-1.

Disadvantages: (1) Additional storage costs for blending, but this is in the interest of aviation, as the capacity of the airport storage facility increases the total storage volume at the airport, thus maintaining a higher safety stock close to consumption. (2) Early transfer of ownership of the JET A-1 to the using airline through a transfer of ownership "into tank" with a return on investment of the inventory, but to a much lesser extent compared to options 1 and 2. Note: This option is currently under further consideration by the JIG.

Option 4 (Buy-back):

Sale of the SAF-SBC to an oil company and subsequent "buy-back of the SAF-SBC" in the blend from this oil company with classical pricing JET A-1 price + SAF-SBC price "free-into-plane", i.e. including all downstream costs included in the location differential plus the costs for blending. The SAF-SBC purchase price is passed on to the airline by the selling oil company in the same amount plus the capital commitment costs for blending.

Advantages: (1) Quality assurance and product liability remain with the selling oil company. (2) Use of transportation route for JET A-1. (3) "Classic" "Free into Plane" pricing for end users. (4) Immediate compensation for delayed or disputed SAF-SBC quantities by delivery of the SAF shortfall as JET A-1 "unblended". Disadvantages: (1) Consent of selling oil company required. (2) Possibly longer transportation distance for SAF-SBC with corresponding impact on the life cycle assessment. (3) Double transfer of ownership with corresponding accounting documentation requirements.

With regard to storage at the airport in an airport fuel depot and removal for aircraft refuelling, there are further differences that need to be taken into account when selecting the least-cost option: In options 1–3, the using airline becomes the owner of the blend "outside the airport perimeter". Consequently, it must enter into a throughput contract for the blended volume with the owner of the airport fuel depot.

For a throughput contract, the airline must always take out and maintain the Into-Plane Liability Insurance, which, in the event of liability damage caused by contaminated or non-compliant JET A-1 or semi-synthetic jet fuel, settles the damage regardless of fault and subsequently takes recourse against the actual causer. The cost of the insurance premium must therefore be added to the total cost of the SAF-SBC. In addition, there are storage throughput charges and aircraft refuelling charges, which would also be included in the "location differential" for conventional JET A-1 "free into-plane" delivery.

In Option 4, there is no need to conclude a storage throughput and aircraft refuelling contract, as the deliveries are owned by the oil company, which in turn already has a throughput contract with the storage operator for its JET A-1 delivery volumes or is entitled to use the storage facility as a co-owner of such a storage facility. However, the selling oil company may object that it does not wish to extend its insurance cover free of charge to the SAF-SBC volumes delivered, even if this does not result in any additional premium expense. The "arbitrage price presumption" or the consumer surplus principle also applies to this situation. In this respect, the alternative liability insurance costs incurred must be added to the total costs of Option 4, at least in terms of calculation.

8.8   FUEL QUALITY MONITORING AND INSPECTION

The quality control procedures currently in place for the production of Jet A-1 and along the supply chain from the refinery to the aircraft relate exclusively to the oil companies and the airlines that are members of IATA. Until now, SAF suppliers – unless they are part of an oil company – have been excluded from participating in these quality assurance procedures. As suppliers of a subcomponent of the JET A-1, however, they must be integrated into the chain of operations, even if this means that insights into the selected technology and the manufacturing process are unavoidable. Neither the oil industry nor the airlines can allow a gap to be created in a safety concept that has been optimised over decades by new suppliers entering the market. As direct or indirect purchasers of SAF, the airlines must insist that all SAF producers are integrated into the safety architecture of the jet fuel supply and, in case of doubt, not purchase the production volumes of an SAF producer until its production, storage and transport are integrated into the planning of the Joint Inspection Group or the IATA IFQP.

The diversity of SAF producers will increase complexity in terms of the number of production facilities and the composition of semi-synthetic jet fuel batches. This may create a quantitative problem in providing the necessary manpower for all planned inspections. If necessary, qualified inspectors will have to be made available as freelance inspectors against appropriate remuneration. In addition, an internationally harmonised database of inspection results from the respective airport fuel depots or aircraft fuelling services would be helpful.

The sooner such procedures and common databases are established, the better for the safety of jet fuel operations at airports and product locations for JET A-1 and SAF.

8.9   LIABILITY REQUIREMENTS AND INSURANCE
COVERAGE FOR SUPPLIERS

8.9.1   AIRPORT STORAGE

The usual terminus of transportation is the airport fuel depot located on the airport grounds. At the beginning of the JET A-1 era, it was common for each oil company to have its own tank farm at an airport. The decisive factor was the construction of underfloor hydrant systems for (time-) efficient refuelling of wide-bodied aircraft at airports with long-haul traffic. Such systems generate high flow rates while simultaneously drawing from multiple hydrants and require appropriately sized storage facilities capable of storing the average consumption of 3–7 days of operation. In such shared storage facilities, the inventories of all users are managed only in terms of quantity and accounting; the identity of the product is lost when it is stored together in an elevated tank. In terms of ownership, there are again differences between North America and the other continents. In Europe, Africa, Asia and Australia, tank farms and underground hydrant systems are owned by consortia in which, for example, the major jet fuel suppliers at an airport have joined together to finance, build and operate an airport tank farm and, if necessary, an underground hydrant system on land leased from the airport company. Oil companies that are not shareholders in such a joint facility may share the use of the facility as the so-called third-party through-putters in return for payment of a throughput fee. Such a right of shared use always exists when the airport operator declares the fuelling facility to be the "central infrastructure" of an airport and thus does not allow any other buildings of this type on its property. In the meantime, airports have also recognised the operation of a fuel farm as a source of income. When building new airports or relocating a fuel depot, airport companies are increasingly investing in the construction of such a tank farm and having it operated by an expert contractor. In the US, in particular, such facilities are jointly owned by consortia of airlines serving an airport or terminal. In contrast to the European ownership model, which provides for equal participation regardless of market share at the airport, the US consortium model at many airports provides for a level of participation and corresponding shareholder rights in proportion to one's own throughput in relation to the total throughput.

In both models, the common storage of jet fuel creates the need for a comprehensive liability regime in the event of a liability claim arising from the operation of the tank farm or caused by the fuel itself when it is handled at the tank farm and delivered to the aircraft by a tanker service.

8.9.2 LIABILITY INSURANCE COVERAGE

The International Indemnity Agreement Consortium requires that each user of an airport fuel storage facility – both shareholders and third-party handlers – maintains liability insurance in the amount of 2 billion US dollars to pay claims for which the policyholder is responsible, regardless of fault, on first demand. The liability provisions of the Tank Storage Agreement provide that in the event of damage, the user on whose behalf the aircraft in question was refuelled shall be liable, irrespective of whether the product stored by the user or the user's fuelling service as vicarious agent caused the damage. The main purpose of this provision is to enable a liability claim arising out of an aircraft accident which can be causally attributed to the fuel or aircraft refuelling operations to be settled quickly and without the need for years of court proceedings to establish the facts of the case. It is understandable that it is then up to the user or his insurance company to seek recourse against the actual party responsible for the damage, to the extent that the damage was caused by a third party and not by the policyholder. In order to provide evidence of possible contamination, tampering or deviation from fuel specifications, reserve samples must be taken from each delivery batch to a tank farm and retained until it is ensured that this delivery batch has left the tank farm and, if applicable, the underground hydrant systems without being mixed with other delivery batches and that all aircraft refuelled with it have completed their flights and landed.

The following example illustrates the operation of the Indemnification Agreement Consortium: (1) Because the Product Owner of JET A-1 is strictly liable and its jet fuel is stored in a commingled storage facility where the physical attribution to the stored quantity of the Product Owner is lost, its liability must be transferred to the jet fuel commingled. (2) The JET A-1 (or semi-synthetic jet fuel) is stored in a bowser. (3) The refuelling attendant refuels an aircraft of the airline "Alpha" with 3 cbm (793 USG) at the airport. The refuelling was done on behalf of the oil company "Charlie". (4) The attendant leaves the parking position of the aircraft of the airline "Alpha" to refuel the next aircraft, which is an aircraft of the airline "Delta" standing on an outside parking position on behalf of the oil company "Foxtrot". On the way to the parking position of the "Delta" aircraft, the driver overlooks a catering truck which has the right of way. There is a vehicle collision, the tank of the bowser is damaged and jet fuel is spilled on the road. (5) The refuelling agent works for the tank farm operator, in which the oil companies "Bravo" and "Foxtrot" are shareholders. (6) The catering company claims damages for the damaged catering vehicle and the airport authority claims environmental damages due to the spillage of JET A-1. At the time of the accident, the Bowser was after the "Alpha" refuelling and before the "Delta" refuelling. (7) For this accident on the airport premises, the physical distance from the location of the accident is measured as a radius to the aircraft of airline "Alpha" and to the aircraft of airline "Delta". (8) If the location of the accident is closer to the "Alpha" aircraft than to the "Delta" aircraft, the "Bravo" oil company shall be liable for the damage. (9) If the location of the accident is closer to the "Delta" aircraft, the "Foxtrot" oil company is liable. (10) In this example, the accident site is closer to the parked "Delta" aircraft. Therefore, the oil company "Foxtrot" settles the claim. It then asks the refuelling company to pay the part of the damage that it has to bear under its professional liability, up to the limit of its cover.

Compliance with the quality regulations for the transport and storage of jet fuel is also subject to regular inspection by the JIG or the IATA IFQP, as insurance cover cannot be maintained without proof of proper compliance with the quality regulations. Consistent compliance with and monitoring of the quality regulations is thus not only a necessary prerequisite for the trade and distribution of aviation fuels, but also the basis for the safety standard achieved today in the handling of JET A-1 and the insurance coverage for all parties involved in the production, transport, storage, blending and distribution of SAF (Buse J., 2016).

Conclusion: The design, construction and commissioning of new plant technology is a challenge in itself. Business plans must be adhered to, the plant must be completed on time and it must deliver the planned production volumes at the required quality in day-to-day operations. This complexity is compounded by the need to create new distribution channels, as the past few years since SAF's first approval have shown how difficult it is to find the right buyers for a new generation of fuels. The pioneering work of the early technology drivers deserves great respect for entering a new market where the vast majority of airlines are cautious and offer little support, even though SAF is obviously the only sustainable option for climate-friendly air transport.

9 Market Implementation Strategies for SAF

After more than 50 years of an oligopolistic supply of JET A-1, SAF is for the first time opening up new options for meeting the needs of the aviation industry. After the selection of alternative raw materials and the technical approval of new conversion processes for the substitution of JET A-1, the question of market entry and market penetration for SAF remains unanswered. Market implementation therefore goes far beyond the primary issue of placing SAF on the market, as only the sustainable achievement of market shares will secure the necessary investments in the supply chain for suppliers and thus make market entry possible. The airlines, as customers, have to answer the question of the optimal supply model for themselves, anticipating future expenses for the compensation of their combustion emissions (as a result of environmental legislation) as well as for higher paraffin prices in the future (as a result of oil price volatility). With the expected long-term approval of SAF neat, market participation will be possible without the involvement of the oil companies, but the expected reactions of the oil industry must be included in the implementation strategies. The sustainable improvement of the greenhouse gas balance and the environmental impact of aviation remains a central component of market implementation.

9.1 SUPPLY ORIENTATION OF OIL COMPANIES

The market implementation of the SAF can be achieved through eight different approaches:

1. Supply-oriented, through appropriate blending offers from the petroleum industry as the current source of supply for airlines.
2. Supply-oriented, through direct neat product offerings from alternative fuel producers whose use as neat products with additives has already been approved.
3. Supply-oriented, through a direct government subsidy of the production cost or sales price, with the subsidy either paid to the producer of the SAF or passed on to the end user through reimbursement.
4. Regulatory, through a legal blending quota that is equally binding on all JET A-1 suppliers and thus creates demand by the oil companies.
5. Regulatory, through a legally mandated blending quota for airlines, which complements or expands the EU-ETS to improve climate protection.
6. Demand-driven, through competitive differentiation of airlines ("green aviation") in terms of the services they offer to passengers.
7. Demand-driven, by exploiting passengers' willingness to pay a premium for a climate-friendly flight.
8. Demand driven by airlines by linking SAF surcharges to passenger privileges.

Basically, there is still the possibility of neutralising the greenhouse gas pollution of the atmosphere caused by air traffic with the help of cross-sectoral compensation measures. To this end, the system of optional CO_2 offsetting, which is currently offered by almost all airlines in the electronic booking process, would have to be expanded. In the event of an international agreement on CO_2 offsetting by the ICAO member states, the current system of a voluntary compensation amount to be paid by the passenger would be replaced by a mandatory contribution to be paid by each passenger – in whatever form – in order to generate the necessary funds for the purchase of CO_2 certificates.

9.1.1 SUPPLY-SIDE MARKET IMPLEMENTATION BY OIL COMPANIES

The international oil companies are currently only moderately involved in the commercialisation of SAF, limiting their activities to selected research topics and market monitoring. Despite approved raw materials and conversion processes, the international oil companies do not currently make their own offers to include SAF in their JET A-1 supply volumes.

This behaviour is economically plausible: The profitability of the upstream business of the oil companies – i.e. the exploration and production of crude oil – exceeds the profits that can be achieved in the downstream business many times over. The price quotations for crude oil and finished products ("specialties") prove the price relationship between upstream and downstream: For example, Brent crude oil is quoted at an average value of USD 81.00/bbl in July 2023 (OilPrice.com, 2023), while JET A-1 is quoted at an average value of USD 101.59/bbl in the same month (JET A-1, IATA Fuel Price Monitor July 2023), which means that the price difference between the raw material and the processed product in the case of JET A-1 is USD 24.59/bbl. Since each ton of SAF displaces an identical amount of JET A-1 (and does not induce additional demand), the complementary production of synthetic diesel and SAF by new suppliers leads to a loss of sales and revenue for existing JET A-1 producers in a static market analysis. If the oil companies expand their own production to include the production of feedstock for SAF, the production costs of the feedstock, including the possibly higher processing losses, must be at least equal to or lower than the production costs of crude oil production, with a differentiation of the production sources according to the level of production costs per production source, whereby the remaining production period of the source and thus its remaining production volume must be taken into consideration. Against this background, the production of feedstocks for SAF is not considered as long as the upstream costs for crude oil are lower than for biofeedstocks, as this leads, ceteris paribus, to an immediate loss of profit for the same co-product output quantity. Under the assumption of a profit maximisation objective, oil companies behave in accordance with the objective if they want to achieve the minimum cost production of the co-products diesel and jet fuel at given commodity prices. Any production of SAF by oil companies therefore implies at least an identical profit margin from the upstream business or from the holistic view of upstream and downstream. For reasons of commercial prudence, a risk premium on the production side is applied to a production process that has not been widely used to date, in accordance with valuation methods customary in the industry.

SAF's additional production of a given quantity of JET A-1 would result in a supply surplus and thus trigger price wars to secure market share and thus lower the sales price for all quantities of JET A-1 sold. Since losses of market share of a commodity product in an oligopoly market can only be corrected by price reductions of the affected firms, which subsequently force all other suppliers to make similar price corrections, the additional revenue from the market share gain is disproportionate to the revenue loss for the entire production due to the ensuing price war, as a result of which the product prices fall until the market shares have almost returned to the initial values before the market share shift or the market participants finally accept the market share shift in order to avoid further revenue losses.

As a result, it can be assumed that, under the given assumptions, the oil companies will not gain any economic advantage from their own production of biofeedstock and SAF, but will suffer losses. Own production of synthetic diesel and SAF by oil companies or purchase from third parties therefore implies new markets as sales channels and the possibility of price differentiation. Co-production requires the fulfilment of these conditions for both synthetic diesel and SAF. A supply-oriented introduction of own production by the mineral oil companies for the jet fuel market with the de facto existing price cap of the JET A-1 product price will therefore take place primarily in Europe in order to fulfil quotas, but will not go beyond this in order to avoid the associated economic risk.

9.2 SUPPLY ORIENTATION OF SAF PRODUCERS

The unit cost of small-scale production (raw material and processing into SAF) is currently higher than the market price for jet fuel. Suppliers are therefore dependent on third-party revenues from the sale of by-products or subsidies to reduce their own production costs to a marketable level. Alternatively, these suppliers would have to generate an additional benefit for the airlines as buyers, which would offset or at least reduce the price disadvantage.

From an economic point of view, any voluntary reduction in the amount of co-products in favour of an increase in the amount of SAF leads to a loss of opportunity in the amount of the difference between the revenue from the co-products and the (negative) revenue from the distribution of SAF. In case of an expected market development, a "start-up loss" would be opportune from a corporate policy point of view if such start-up losses could be amortised in the future through competitive advantages in the market as a "first mover" with a higher sales volume. In view of the amortisation risk, the current producers of synthetic fuels prefer the exclusive distribution of co-products and refrain from producing SAF or limit the production of SAF to the minimum SAF production in terms of costs.

Since the majority of potential suppliers for future SAF production already produce a significant amount of synthetic diesel due to tax incentives (and the legal obligation to blend) (and thus already benefit from the economies of scale of industrial production), the proposal discussed by aircraft manufacturer Boeing to blend up synthetic diesel in JET A-1 could have laid the foundation for entry into SAF production. However, no progress has been reported on this concept due to ongoing concerns of engine OEMs regarding the cold flow properties of such synthetic diesel blend. Instead, the principal concept has now been integrated into the co-processing option during the refinery processing of crude oil. From the perspective of the oil companies, they will demand at least the JET A-1 price for any co-processed blend, but possibly a higher price, arguing that there is no blending requirement for co-processing. Conversely, from the airlines' point of view, they will insist on a price based on marginal costs, since it is a question of expanding the existing sales channel for co-processed feedstocks, even if the above-mentioned volume substitution of JET A-1 takes place, and the suppliers will only offer the co-processed blend if it does not worsen their profit margin. However, the blending of a product with a different specification could require changes to the additives in the neat JET A-1, so that the co-processing will have to be subject to special quality assurance and cannot simply be added to the corresponding JET A-1 on a case-by-case basis.

In any case, the co-processing of suitable oils and fats will have a positive impact on feedstock production, as ideally the additional global demand for co-processing components could be as high as 2.5 million tons per year. The concentration of air traffic in the US, Western Europe and Southeast Asia would justify the construction of additional refining capacity with an annual output of 800,000 tons in each of these markets, provided that the "aviation price" for co-processed JET A-1 is at the JET A-1 price level.

9.3   SUPPLY DRIVEN BY GOVERNMENT SUBSIDIES AND INCENTIVES

This option represents a hybrid between the purely market-based supply options and the government regulation options. In this variant, the additional costs above the JET A-1 reference price level are compensated by direct subsidies. On the producer side, such a subsidy is used as an "OPEX subsidy" to reduce production costs. As an entry option, a government subsidy can certainly have a market stimulating effect. However, there is a risk that a market model will emerge that is designed for permanent government subsidisation, and that the subsidies will not serve as a catalyst for a self-sustaining market model that creates a long-term equilibrium between supply and demand, in which economies of scale lead to successive reductions in production costs that make government support unnecessary over time. There is no doubt that subsidies increase airlines' willingness to buy if the SAF price level is reduced to the point where competitiveness with the JET A-1 is achieved, albeit artificially. However, the subsidy period must be used by the market participants to achieve the economic viability of SAF production under the subsidy umbrella. Otherwise, the airlines' willingness to buy will end with the end of the subsidy.

This issue touches on the current subsidisation of American SAF production through high investment grants on the manufacturer side and the possibility for airlines to use the RIN tax credits they receive to compensate for the price premium for SAF compared to fossil JET A-1.

In the competitive situation between HEFA and ATJ jet fuel and PtL jet fuel as RFNBO with significantly higher production costs compared to the biomass-based technology paths, an OPEX subsidy for PtL jet fuel can promote the marketability of PtL in Europe, which would otherwise have no sales opportunities due to prohibitively high production costs. From a strategic point of view, an OPEX subsidy in the start-up phase would further support the maturation of the PtL technology, but it cannot be provided as a permanent subsidy in order not to distort competition.

9.4 REGULATORY REQUIREMENTS THROUGH PRODUCER MANDATES OR QUOTAS SET BY GOVERNMENTS

In this model, JET A-1 sellers are required by regulation to blend a predetermined amount of SAF into their product. Since any blending will always be limited to the product characteristics of the particular batch of JET A-1, a legal blending quota can only be set at a level that ensures compliance with the specification limits. As is common practice in other sectors, non-compliance with the blending obligation would trigger compensation payments by the producers to the tax authorities of the respective states, the amount of which would have to be at least equal to the price difference between the JET A-1 price and the SAF production costs in order to create an incentive to produce. If an oil company does not have its own production facilities to produce SAF, it would have to purchase the required blending volumes from other producers or, alternatively, pay compensation payments until it has established its own production capacity, if this is granted by the governments. For their part, companies that already produce synthetic fuels will increasingly act as demanders in the feedstock markets. The resulting temporary market imbalance will lead to an increase in feedstock prices in the short term, thus stimulating the cultivation of feedstocks or the increased use of recyclable waste materials. Under competitive conditions, those raw material processing alternatives that can respond quickly to changes in demand and realise economies of scale in expanding production will prevail in the context of regulation (Neste, 2020).

Regulated airlines have limited ability to influence market developments. Assuming the continuation of the existing "free-into-plane pricing model" with formula prices linked to spot market quotations (for JET A-1 or for feedstock), the blending quota leads to a uniform price movement for JET A-1. Theoretically, the airlines "save" the purchase of emission certificates in emissions trading for the SAF quantities purchased as "blending components", at least to the extent of the CO_2 reduction eligible for credit. In such a model, the seller would have to issue the airlines with corresponding credit certificates for the SAF blending quantity or, in the case of non-blending of SAF and corresponding compensation, reimburse its aviation customers for the value of the additional emission certificates they now have to purchase.

In light of the ICAO member states' negotiations on a global emissions trading system for international aviation (CORSIA), it seems unlikely that CO_2 emissions trading in the EU will be replaced by a global blending obligation on the part of the mineral oil industry for aviation fuels in the period up to 2027, as the necessary raw material quantities and production capacities are not available at present and production will take 5–7 years to set up. However, the European trilogue compromise on the amendment of the EU-RED regarding the introduction of blending quotas provides for a continuously increasing SAF quota from 2% in 2025 to 70% in 2050 (which is not feasible according to the current ASTM certification situation). In addition, the EU Commission is pursuing the goal of reducing the use of cultivated biomass for energy purposes. In principle, a blending quota for SAF should also be achievable for the market participants involved. In this respect, a neutral survey of the available raw material potentials as well as the suitable waste and residue quantities for cascade use in SAF production would be a prerequisite for setting blending quotas. The available research results need to be compared and harmonised with regard to their survey methodology. Given the coordination and time required for global harmonisation, blending mandates can only be implemented promptly in those countries that can undoubtedly provide the required volumes of feedstocks – preferably without competing with existing supply chains for ethanol for gasification fuel and for FAME biodiesel as a blending component for diesel fuel.

From the airlines' perspective, the oil industry's blending mandate leads to the market implementation of SAFs, but without any real opportunity to control the associated price changes for aviation fuels. On the other hand, the ease of sharing the oil companies' storage and transportation facilities and the continuation of the oil industry's commitment to meet demand and supply in the existing contract model between oil companies and airlines remain advantageous.

9.5   REGULATORY REQUIREMENTS THROUGH SAF PURCHASE MANDATES OR GOVERNMENT QUOTAS

A legal obligation to purchase SAF volumes in the amount of a percentage of the total consumption of each airline would not be directed at the producers or distributors, but at the end users as the controlling subject: Compared to a blending obligation for manufacturers, a use obligation for aviation offers several options for action and thus at least the possibility of influencing fuel costs:

1. The purchase of SAF can be contractually decoupled from the purchase of JET A-1. In this case, however, the airlines would have to purchase SAF individually or jointly in the form of a purchasing group or use the services of a third party.
2. The purchase of SAF may also be carried out individually or jointly by a "purchasing group" of several airlines, thus exploiting synergy effects which an individual airline could not realise on its own due to the volume involved, e.g. joint transport and joint storage of SAF quantities. Provided that the size of the association does not lead to a dominant position and the air fares for the end consumer are not influenced by such alliances, the joint purchase of consumables is considered to be unobjectionable under competition law.
3. Airlines may enter into long-term purchase commitments with SAF manufacturers and thereby influence the product price.

4. In addition, airlines may, in principle, agree on price formulas other than the usual formula price linked to spot market quotations.

5. The purchase of SAF quantities can be combined with the purchase of emission certificates. This applies in particular to the SAF supply chain for the cultivation of vegetable oils in developing countries.

6. Airlines may also fulfil their blending obligation – if any – by producing their own SAFs through their own investments in the process chain or through proportionate investments in raw material production and processing. Insofar as the vertical integration of the process chain into the own supply leads to a higher return than the return from the core business of offering air travel and air cargo transportation, an own investment in the production of SAF leads to an improvement in the average return of the investing airline.

In any case, direct purchase of SAF by airlines (with or without quota) would stimulate the market and lead to demand-driven production. This would allow airlines to focus on the raw materials and production processes best suited to their needs and to create a cost-optimised supply chain for at least part of their SAF requirements, which would ensure a cost-effective supply. From a risk diversification perspective, airlines can participate in such SAF supply chains on a reciprocal basis (e.g. within airline alliances) and thus develop cost advantages for their flight operations.

9.6 DEMAND DRIVEN BY AIRLINE COMPETITION DIFFERENTIATION

Most of the major European airlines emerged from former state-owned ("national") airlines.

As part of the supply-side economic policies of the 1980s and 1990s, European air transport was also deregulated. Nevertheless, the former state-owned airlines still play the role of "national carriers" today, tailoring their flight schedules to the needs of their growing home markets. In this respect, the customer base of the domestic market exerts a decisive influence on the network offer and product design.

Today, large corporations are viewed critically by socially relevant groups in OECD countries, and socio-politically compliant behaviour is assumed. A company's social code of conduct is referred to as Corporate Social Responsibility (CSR) and includes environmental policy measures.

As the environmental preferences of passengers – especially frequent travellers – increase, the environmentally friendly behaviour of the preferred airline increases long-term customer loyalty. In this context, the booking behaviour of companies that conclude corporate travel contracts with airlines for the flights of their employees is becoming increasingly important: Especially companies that are active in the consumer retail business ("B2C") and/or have set environmental standards for themselves demand not only volume-based pricing for their business trips ("corporate travel") but also "green activities" of the contracting airline (compensation of Scope

3 emissions). Due to the economic importance of corporate travel contracts for overall revenues, the major European airlines cannot avoid providing evidence of their own environmental activities. Similarly, there is growing pressure in the air cargo business to demonstrate to shippers that air cargo transportation is environmentally friendly by reducing CO_2 emissions.

From the point of view of securing long-term revenues from high-revenue corporate customers, the introduction of SAF offers airlines a good opportunity to differentiate themselves from competitors who are less committed to environmental policy. However, market segmentation in sales must be taken into account: Companies that generate their revenues in a price-sensitive customer segment with only limited environmental orientation can understandably not expect a corresponding benefit in every case. Companies with a professional purchasing organisation, on the other hand, use modern evaluation methods when comparing offers, which also include qualitative factors in the award decision and thus reflect the added value of a climate-friendly flight in a price-congruent manner ("value-based purchasing").

With the start of the long-term use of SAF, environmental organisations will express their positions on the use of biofuels. Airlines therefore run the risk of achieving exactly the opposite in the external perception of their customers, despite their environmentally friendly intentions. If the arguments of the environmental organisations cannot be clearly rejected as incorrect in a specific case, there is at least the risk that corporate customers will distance themselves from the airline (and thus outsource their air travel or cargo shipments to competitors) in order to avoid public pressure against the use of SAF or to prevent damage to the company's image. The criticism of the environmental organisations is mainly directed at the cultivation of raw materials and the related issues of (1) land use change (dLUC and ILUC), (2) food security, (3) water use and non-impairment of subsequent water users, (4) adequate consideration of local and regional interests (especially existing land use and grazing rights of the local population) and (5) nature conservation, as well as (6) adequate participation of local land users, communal bodies and political representatives in land development and the establishment of cultivated areas.

In view of the very extensive catalogue of requirements for environmentally and socially compatible production of raw materials, the question arises as to whether airlines should procure certified SAF from sources unknown to them, or whether a more extensive involvement in the process chain is required in order to avoid possible damage to their reputation and to be able to influence conflict situations as they arise, irrespective of whether certification already exists. It must be taken into account that the economic consequences of a damaged reputation can exceed the company's own control and monitoring efforts many times over. German environmental organisations do not consider the certification of the cultivation of raw materials by the certification systems accredited in the EU to be sufficient. In addition, certification is only a snapshot in time and can only inadequately cover, for example, occupational health and safety and compliance with company regulations in day-to-day operations.

The approach to achieve competitive differentiation through the use of SAF therefore requires careful preparation and embedding in a holistic CSR strategy that is perceived as comprehensible and credible by the outside world and thus complements the corporate values and culture.

9.7 DEMAND DRIVEN BY AIRLINES THROUGH SKIMMING OF PASSENGERS/CARGO BY SAF SURCHARGE

If there is sufficient acceptance, market penetration and customer demand, the implementation of BioSPK can also be financed by the passengers by passing on the additional costs to the end customer through pricing or special surcharges. However, the design of such diffusion models must take into account the estimated time required for the necessary change in customer preferences (Figure 9.1).

Diffusion is the length of time it takes for an innovation to permeate a social system. This is especially true for long-term customer loyalty. Put simply, companies strive for deeper customer loyalty at a higher price as a marketing goal. This paradox can be implemented in a way that increases revenues, especially in the case of intangible values. In other words, the need for an environmentally friendly flight can be seen as an up-sell potential and maximised in terms of revenue. The additional benefit of the surcharge must be recognisable and desirable to the passenger in order to make the countervalue of the emission reduction visible. Experience with voluntary carbon offsetting solutions shows that only a small percentage of self-bookers take advantage of this option, partly because the environmental benefit is only recognisable in the abstract and has no direct connection to the actual flight. In order to ensure that the additional revenue remains available to finance SAF volumes and is not absorbed by the additional costs of "green" privileges, the privileges to be offered

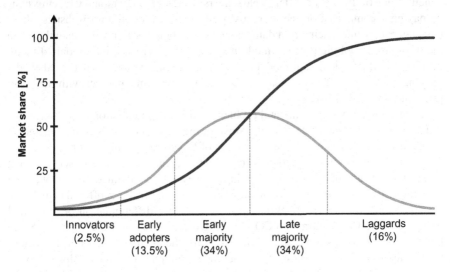

FIGURE 9.1 Adoption model of market penetration according to Rogers.

Source: Wikipedia (2023) – Diffusion of Innovations (prepared by author).

to passengers in return may be only or mainly intangible in nature, e.g. based on travel status and related preferential treatment.

In a specific case, the airline can assure the passenger that it will use the amount of a voluntary surcharge paid – after deducting purchase and administrative costs – to cover the additional costs of SAF (compared to comparable costs for the same amount of JET A-1). Proof of the use of the funds can be certified by the auditors in the annual financial statements and communicated in the report on the results for the financial year. Such compensation schemes may be used by the airline to supplement its CSR policy for accompanying measures, e.g. in the cultivation of raw materials in developing countries.

9.8 DEMAND DRIVEN BY AIRLINES BY LINKING SAF SURCHARGES TO PASSENGER PRIVILEGES

The granting of privileges in return for CO_2 offsetting through the purchase of SAF in relation to flight distance applies to both purely short-haul and long-haul flights. For the passenger, the relevant differentiation criteria are primarily the airfare, the total travel time and the time of departure and arrival, as well as travel comfort, assuming daily service on the respective routes. In addition to pricing, competition focuses on non-tariff differentiators, with certain basic services being offered equally by all airlines for each class of travel because passengers have come to expect them as "class-typical features". To the extent that comfort components are included in the prices of all carriers, the development of USPs to differentiate from competitors is a marketing task that relates not only to the passenger's personal travel comfort but also to his or her values and attitudes. To gain passenger preference for an airline, the goal is to influence the selection decision by offering various intangible additional benefits. One of these is the offer to passengers to offset their personal CO_2 emissions by paying a compensation fee. For this purpose, airlines maintain websites where passengers can enter their travel data and have the compensation value calculated. In the web-based booking process, modern systems transfer the booking data directly into a CO_2 offset calculation model and display the offset value. If accepted by the passenger, the CO_2 offset amount is collected by the airline together with the ticket price and invested in SAF.

In this context, the fundamental problem of price differentiation in passenger air transport must be taken into account, since in the so-called "revenue management" of the airlines, passengers in the same class of transport pay different prices for the same service. Price differentiation is achieved by segmenting a class of service, such as economy class, into different booking classes with predefined seat quotas and booking windows. In this environment, where the same service is offered at different prices – depending on the booking time – the further differentiation of individual passengers into CO_2-neutral and CO_2-emitting passengers leads to a dilemma: Since both passenger groups travel in the same aircraft, the environmental benefit of their CO_2 compensation – regardless of whether their compensation amount flows into the purchase of SAF or is used to support other environmental projects with CO_2 reduction – is not perceivable for the CO_2-compensating passenger. On the other hand, the differentiation from competitors is precisely the appreciation of environmental protection as an expression of socially desirable behaviour. In terms of content, this topic belongs to the area of customer loyalty and the customer loyalty programmes of today's airlines. One way of differentiating from the competition is therefore to further differentiate customer loyalty programmes, which can design context-related offers for particularly environmentally conscious passengers. A number of benefits are available as intangible privileges that are of value to the passenger, but generate little or no cost to the airline: These could include "sustainability ambassador" status, which allows passengers to participate in gate pre-boarding, priority on waiting lists for rebooking on other flights that are already fully booked, priority booking in rows with a wider seat pitch, credit for a passenger's "ancillary income" on food, beverages, duty-free purchases, additional baggage and CO_2 compensation, additional baggage, and CO_2 offsetting for customer status or delayed expiration of "green" miles in customer loyalty programmes, separate purchase option for travel accessories available only to CO_2 offsetting customers, or lower surcharges for upgrades to a higher class on long-haul flights based on available seats in Premium Economy or Business Class (Table 9.1).

Overall, airlines should strive for a better image, because "Corporate Social Responsibility" is more than a slogan; it is the social behaviour of a large company that accepts its responsibility for all consequences related to its performance, production and professional business activities.

9.9 REPLACING SAF WITH OFFSETTING THROUGH ALLOWANCES

IATA's alternative of neutralising the greenhouse gas impact in the balance sheet by purchasing more carbon credits and thus providing an equivalent option instead of using SAF is based on the assumption that the increase in production capacity for SAF and the resulting availability of "green" synthetic jet fuel quantities will not be congruent with the development of demand in aviation for climate-friendly aviation fuels. In this respect, the alternative course of action of "compensation payments" serves as an argument for closing the SAF supply gap.

However, IATA's approach is not without risks: as long as CO_2 compensation through the purchase of certificates ("cap & trade" – especially through the purchase of CDM certificates) can be presented as cheaper than the purchase of SAF quantities in a price comparison with JET A-1, airlines will prefer the option of purchasing certificates, which is available to them on short notice and cheaper at the same time. In parallel, this will reduce the short-term demand for SAF, which is necessary to initiate the long-term production build-up. Understandably, the necessary investments in an SAF supply chain will only be made if the investment calculation with the assumed demand quantities ensures the necessary amortisation of the invested capital and a corresponding minimum return. Any weakening of demand will inevitably lead to a delay in the investment decision if the revenue assumptions are undermined by contrary behaviour on the part of the airlines. In this context, there is a danger that the alternative conceived as a "gap-closing" measure will in fact develop into a barrier to market entry for SAF systems, and that the price advantages of SAF production that can be expected in the long term will not materialise under the given conditions due to a lack of capacity expansion!

TABLE 9.1

Impact of SAF Success Factors on Airline Strategies Assessment by Author Based on Typical Strategy Rankings in the Transport Sector

SAF Implementing options	Balance	Success Factors and Strategies for Airlines				
		Profit Maximising	Environmental Contribution	Cost Leadership	Price Differentiation	Product Differentiation
		+2	+15	−6	+6	+6
1 SAF as blend product	−1	−	+	− −	0	+
2 SAF as neat product	+5	+	+++	− − −	++	++
3 Government subsidies	+11	++	++	+++	++	++
4 Blending mandate (suppliers)	−9	− − −	++	− −	− − −	− − −
5 User mandate (airlines)	−4	− −	++	−	−	− −
6 Competition differentiation	+10	++	++	0	+++	+++
7 Surcharge skimming	+11	+++	+++	−	+++	+++

Scaling and ranking of positive impacts, +++, ++, +; ranking of negative impacts, − − −, − −, −; neutral, 0 (prepared by author).

Conclusion: Airlines and gas stations share the same problem – they are looking for ways to retain customers in a commodity market where price is the deciding factor for many customer segments. In the same way that a price difference of a few cents per gallon leads people to choose the cheaper gas station, leisure travellers in particular will choose a different airline and accept longer travel times, even if the price advantage is marginal. Business travellers are more likely to look at flight times and departure and arrival times, and accept higher prices if it increases their productivity and reduces their travel stress. Between these extremes, SAF must become part of an airline's marketing strategy to capture the attention of passengers.

10 Summary and Outlook

In 2009 – the year the first SAF-SBC was qualified – the jet fuel consumption of civil aviation was 66 billion USG or 200 million tons. By 2019, jet fuel consumption had grown to 95 billion USG or 288 million metric tons (Statista, 2023). In just 10 years, global jet fuel demand increased by 29 billion USG or 44%, while specific fuel consumption decreased due to the partial replacement of older aircraft with more efficient successor models. If the fleet had grown with the aircraft technology of 2009, the increase in jet fuel consumption would have been even higher. In 2019, SAF accounted for less than 500,000 metric tons of global jet fuel production, or less than 0.2% of global jet fuel demand.

The principle of "carbon-neutral growth" only applies to aviation from 2020. In the ICAO CORSIA system, the share of international flights of the 288 million tons in 2019 will initially remain tax-free. For intra-European flights, which consume about 45 million tons of kerosene, a significant portion of consumption will remain tax-free until 2026, when it will be fully subject to emissions trading under the EU-ETS. In the US, the LCFS is leading to a steady reduction in the carbon intensity of JET A-1, but further steps are needed to reach the target.

In fact, with the end of the Covid19 pandemic, international civil aviation is in the midst of its greatest challenge since the introduction of the jet age in the 1960s: Aviation emissions must be permanently reduced in the very near future. Attempts to achieve this goal through market-based measures will not be successful given the magnitude of the problem, but they will have to make a limited contribution to the overall success. Despite having been allowed to do so for 14 years, the airlines have not yet been able to come to terms with changing the type of fuel they use, even though this is the only option for the future that will visibly and demonstrably reduce aviation emissions. Policymakers have made it easy for airlines to hide a serious problem by pointing to fuel-efficient aircraft and fuel savings through weight reduction and operational optimisation. The record of a 44% increase in jet fuel consumption in just 10 years shows the whole truth: doing nothing is not the solution to the problem. The facts are forcing the industry to embark on a transformation process that will likely take more than 20 years for aviation to truly achieve low emission operations. In order to achieve this, JET A-1 will have to give way to a substitute product that will be significantly more expensive to manufacture and will only be available at significantly higher prices, while maintaining comparable product characteristics. The opinion-forming airlines of IATA have at best failed, at worst ignored, to take advantage of the unique opportunity presented by the change in fuel type to eliminate the speculative element in formula pricing through their own fuel supply activities and thus create a new investment climate in which new suppliers with fresh ideas can initiate an SAF fuel supply as reliable as that successfully implemented by the oil companies more than 60 years ago.

DOI: 10.1201/9781003440109-10

The awareness that we are in a "race against time" is now dawning on the public. The "Fridays for the Future" movement of schoolchildren had already taken the lead in climate protection discussions in 2019 and massively increased the pressure on governments when the Covid19 pandemic paralysed public life and the air transport system. Today, in the summer of 2023, climate change is back on the agenda. According to media reports, global air traffic in 2024 will be back to where it was in 2019 at the beginning of the pandemic – with rising emissions. Meanwhile, the scientific community has taken advantage of the forced pause of the Covid19 pandemic to address the issue of "non-CO_2 emissions from aviation". The message is clear: contrails and other greenhouse gases contribute more to global warming than CO_2 emissions. Meanwhile, airline marketing departments are running out of environmental talking points: Airplanes are already cramped, inflight services have been reduced and planting more trees as an offset measure will not help, because the earth is too small to grow all the trees the airline industry would like to plant.

Before things go the way they shouldn't, it's time to ask what the long-term solution for aviation looks like when politicians think only in terms of the time horizon until the next election.

This book should have provided hints and suggestions. But in case the insight has been lost in the multitude of factors to be considered, here is the final summary:

1. If there is only one environment, environmental standards must be globally harmonised or at least mutually recognised as equivalent. This is the essential prerequisite for global investment in a global SAF market.
2. Political actors need to set priorities: If emissions reduction is to be a priority, all restrictive laws and regulations that undermine or delay the goal must be suspended for a period of time to be defined in order to accelerate the transformation process. If success is irreversible, restrictive laws can be reactivated step by step.
3. It is time for a round table of airlines, oil industry and SAF producers to shape the jet fuel supply transformation process and create a win-win situation for all stakeholders: If airline CEOs are personally committed to these goals, the capital market will believe in the seriousness of the transformation process.
4. It is never too late to start today. Let's preserve the freedom to travel that air travel offers, bringing us closer to other people, cultures and landscapes, and broadening our personal horizons.

Glossary of Terms

Air Carrier: A term used in the Warsaw Convention: An air carrier is an entity that, as a contracting party, concludes a contract of carriage with a passenger or consignor for the transportation of passengers and/or cargo and is liable for such transportation, subject to the limitations of liability set forth in the Warsaw Convention. In the case of codeshare flights, the passenger travels under the Carrier's flight numbers (even if the operational performance of the flight is carried out by a foreign airline). In this case, third-party airlines act as agents of the Carrier. The Carrier's fare conditions apply. In the case of interline flights with third-party airlines, the Carrier acts as agent for the third-party airline(s) for individual routes. In the event of service disruptions, the terms and conditions of the airline responsible for the disruption may apply.

Backwardation: A situation in the oil market of falling prices in which forward contract prices for future delivery are lower than the expected spot market prices at the time of delivery. The opposite situation is called "contango").

Belly: The underfloor cargo hold of a passenger aircraft.

Block Time: Parked aircraft are secured against uncontrolled movements by blocks fixing the wheels of the undercarriage. Once an aircraft leaves its parking position, the blocks are removed to give way for wheels to turn. The aircraft proceeds to the runway and takes off. Upon landing, the aircraft rolls to its final parking position and, upon arrival, the wheels are secured again by blocks. The "block time" is the time from block removal when commencing a flight until blocks are being placed again upon completion of a flight. Flight schedules usually refer to the block time as the departure and arrival time of an aircraft. Flight time is the time when the aircraft becomes airborne at take-off until landing on the runway at the destination airport. Based on the effective departure time ("off blocks"), the airline's flight management system sends an automated message to the arrival airport with the estimated "on block time" of the aircraft.

Codeshare Flights: A flight with multiple flight numbers. In addition to the own flight number of the airline operating the flight ("carrier"), the flight number(s) of the airlines that may book seats on this flight for their own passengers and that designate this flight in their flight plan as part of a multi-sector flight. Codeshare flights are usually inbound or outbound flights that are coordinated with the departure and arrival times of long-haul flights and thus involve short transit times.

Diffusion of Innovation: This theory reflects how fast, when and why new technologies are adopted by consumers and markets. The theory was originally developed in the 19th century, but became known and popular through Everett Rogers' book *"Diffusion of Innovations"*, first published in 1962. According to Rogers, diffusion is the process by which an innovation becomes known over time in a social system or the relevant group of consumers.

To some extent, the theory combines elements of the life cycle concept with the dimension of time in customer adoption of a new product or technology.

EMCS: EMCS stands for Excise Movement and Control System and is a European database that monitors all movements of controlled products within the borders of the 27 EU member states. As jet fuel is exempt from excise duty, all movements of jet fuel from refineries or imports are monitored and deliveries are tracked. The system provides a "closed shop" situation for quantities of jet fuel accounted for and delivered. Under EMCS, it is impossible to use undeclared product quantities for other purposes, for example to avoid downstream taxation.

European Green Deal: The European Green Deal is a concept presented by the European Commission on 11 December 2019 with the aim of reducing net greenhouse gas emissions in the EU27 to zero by 2050, making it the first "continent" to become climate neutral.

Food Security: FAO definition of food security and its dimensions:

Food security exists when all people, at all times, have physical and economic access to sufficient, safe and nutritious food that meets their dietary needs and food preferences for an active and healthy life. (World Food Summit, Rome 1996)

Food security has four dimensions:

1. Food availability: The availability of sufficient quantities of food in adequate quality, supplied by domestic production or imports (including food aid).

2. Access to food: The access of individuals to adequate resources (entitlements) to acquire adequate food for a nutritious diet. Entitlements are defined as the set of all commodity bundles over which a person can establish control, given the legal, political, economic and social arrangements of the community in which they live (including traditional rights such as access to common resources). Utilisation: The use of food through adequate nutrition, clean water, sanitation and health care to achieve a state of nutritional well-being in which all physiological needs are met. It emphasises the importance of non-food inputs in food security.

3. Food stability: To be food secure, a population, household or individual must have access to adequate food at all times. They should not be at risk of losing access to food as a result of sudden shocks (such as an economic or climatic crisis) or cyclical events (such as seasonal food insecurity). The concept of stability can therefore refer to both the availability and access dimensions of food security.

4. Food utilisation: People make appropriate use of food through storage and processing practices. They have adequate knowledge of the nutritional, health, hygiene and socio-cultural parameters of food.

FORDEC: FORDEC is an acronym that stands for Facts, Options, Risks, Decision, Execution and Control. FORDEC was developed by the German DLR (German Aerospace Center) in 1994 as part of the so-called Crew Resource Management development. The methodology calls for a condensed and concentrated analysis of a problem that occurs during a flight and requires immediate action by the pilot. The goal is to gain time for decision-making and to find the most appropriate solution to the problem. The "facts" of the problem are noted and the "options" for solving the problem are recorded. For each option, the "risks" and "opportunities" are weighted. Based on this information, a "decision" is made for the best option. Then, the "execution" is the action initiated to eliminate or mitigate the problem. At the end of the flight, "controlling" evaluates how well the problem was handled and what lessons were learned that will assist other pilots in similar situations.

Fuel-to-carry: The "fuel-to-carry" effect describes the additional consumption of aviation fuel relative to the distance flown due to the weight of aviation fuel carried over the distance flown.

In constant cruising flight, the aircraft becomes lighter due to the successive consumption of the fuel carried and can climb to higher altitudes with less drag (due to the decreasing density of the ambient air) in relation to the weight loss due to the fuel consumed. For the purposes of the following discussion, however, the fuel effect of the lower frictional resistance due to the change in altitude during cruising flight is of secondary importance.

However, depending on the length of the flight, additional fuel must be carried in addition to the regular amount of fuel to carry the weight of the fuel over the distance. As the distance of a flight increases, this effect becomes more pronounced because the transportation of the remaining fuel causes additional consumption due to the weight of the remaining fuel, which in turn increases the total consumption of the aircraft over the selected distance. This secondary effect is called Fuel-To-Carry (FTC).

For example, for an A380 route length of 3,000 NM (equivalent to 5,560 km), the additional fuel consumption is 7,034 kg. At 4,500 NM (8,330 km), the additional consumption is 16,884 kg, and at 6,000 NM (11,110 km), a total of 30,955 kg of aviation fuel must be refuelled as FTC. The total amount of aviation fuel for this distance is 162,097 kg. The FTC share for the 6,000 NM distance is calculated as 19.1% of the total consumption (data evaluation by the author).

Interline Agreements and Interline Flights: Agreement between two airlines to mutually accept tickets and reservations in each other's reservation systems. The company issuing the ticket (the passenger's contractual "carrier") books the passenger in individual segments on flights operated by the interline partner and transfers a contractually agreed amount for the passenger's carriage on a third airline to that company, with only the balance of the reciprocal bookings being settled. The passenger on an interline ticket changes airlines during the flight. Baggage is checked through and the passenger receives boarding passes for all individual flights, including those of the foreign airline(s), at the beginning of the trip. The number of interline passengers on interline flights may be subject to booking restrictions, such as a limit on the number of interline passengers per flight.

Into-plane Insurance: Special liability insurance for suppliers of aviation fuel with a limit of indemnity of USD 2.0 billion for the settlement of liability claims in connection with the refuelling of aircraft and the quality of aviation fuel, irrespective of fault. The special feature of the coverage is the no-fault settlement by the insurance company of the respective supplier, even if the actual cause of the loss is presumed to lie with a third party (e.g. with another oil company as a result of the joint storage of aviation fuels, where the identity of the property is inevitably lost). In the case of a subsequent recourse, the defence of the amount of the settlement by the party causing the damage is excluded.

Joint Guidelines (JG): Voluntary quality and safety standardisation of the mineral oil industry for the handling of aviation fuels. The regulations are only available to members. The organisation of operations in accordance with the described quality and safety requirements is a prerequisite for the conclusion of the required public liability insurance and the so-called "Into-plane Insurance".

Joint Inspection Group (JIG): Voluntary working group of safety representatives of international oil companies to establish industry standards (Joint Guidelines) for the operation of tank terminals and the transport and storage of aviation fuels. The unique feature of the JIG is the cross-competitive, periodic inspection of depots and logistics facilities by inspectors from different companies.

Life Cycle Analysis: A Life Cycle Analysis is a systematic analysis of the potential environmental impacts and energy balance of products throughout their entire life cycle. Before each analysis, the system boundaries are defined, as different boundaries make sense depending on the product and the objective of the analysis (Wikipedia, 2023). For example, for the production of SAF, emissions from the cultivation of raw materials (including the use of chemical fertilisers) up to the "first collection point" are covered, as well as transport emissions from this collection point to the pre-treatment plant or directly to the SAF production plant, including process emissions during the production of SAF and emissions during storage, blending, transport and storage at the airport up to the point of aircraft refuelling. The emissions of the individual segments are either collected on a product-specific basis or default values are included in the emissions calculation as accepted reference values.

Maximum Take-Off Weight (MTOW): The MTOW describes the maximum weight of an aircraft for a safe departure with an unexpected engine failure during take-off. The MTOW is dependent on the weight of the aircraft, fuel weight and payload weight (passengers, baggage and cargo) in relation to the available runway length and the ambient airport temperature at the time of departure. Upon acceleration of the aircraft on the runway, V1 is the speed at which a take-off can be aborted and the aircraft will stop ahead of the end of the runway. Beyond V1, the aircraft has to fly as the remaining runway is too short for a breaking manoeuvre. VR is the speed when the pilot in command rotates the aircraft by lifting the nose for building up lift on the wings. V2 is the safe minimum speed for the aircraft if an engine failure occurs. The MTOW is calculated prior to each flight.

O&D – Transportation: Analysis of passenger traffic by origin and destination. The passenger volume of a departure point provides information on the size of the source market ("catchment area") and the preferred destinations from this source market. Depending on the traffic volume, O&D connections are linked by a non-stop flight or by connecting flights.

Point-to-Point Traffic: Number of passengers travelling between two airports on a single non-stop or connecting flight.

Posted Airfield Price: Published standard retail price for jet fuel at an airport for customers without a supply contract. Typically used only for small delivery volumes and ad hoc single flights, as the price is usually unattractive due to the inclusion of holding costs.

Propulsion: Today's commercial aircraft are powered by jet turbines or turboprop engines. In both cases, air is drawn in, compressed and then burned in a combustion chamber. The hot combustion gas exits through a series of blades that are rotated by the speed of the exhaust gas. A shaft transmits the rotary motion to the engine fan or propeller. Combustion of the kerosene–air mixture is constant.

REACH Certification: REACH is a European regulation and stands for Registration, Evaluation, Authorisation and Restriction of Chemicals. The overall objective of REACH is to provide a high level of protection for human health and the environment from the use of chemicals. REACH also allows for the free movement of substances in the EU market, such as jet fuel or SAF SBCs.

Spot Market: The place where supply and demand for spot or cash transactions meet. Originally established for foreign exchange and securities transactions, spot markets are also used for the sale of standardised commodities in the sense of a commodity futures market.

Third-Party Throughputters: Holder of a throughput contract with a jet fuel depot operator (third-party operator, i.e., not a shareholder of an airport jet fuel depot). The right of use relates to the storage of jet fuel at the depot and its delivery to the customers of the third-party throughputter.

Trilogue: Trilogue is the name of EU internal consultations between the three institutions: EU Commission (comparable to national Government), EU Parliament (comparable to a national parliament or House of Representatives) and the EU Council (members are the Prime Ministers or Chancellors of the 27 EU member states). In case of disputes between the three EU governing parties, the trilogy negotiations are staffed with representatives of the three institutions jointly seeking for a compromise. Any compromise must be accepted finally by the EU Parliament by the majority of votes.

Yield Management: An American term for revenue planning using dynamic pricing to achieve maximum revenue from price and capacity utilisation. Used for time-based services such as room rentals, car rentals, air and rail travel.

Bibliography

Air France/KLM (2023): Financial results. https://www.airfranceklm.com/en/finance/financial-results. 2022.

Argus Media (2020): European UCO buyers wary of contaminated imports. https://www.argusmedia.com/en/news/2127050-european-uco-buyers-wary-of-contaminated-imports 28. July 2020.

Argus Media (2022): Swedish companies to produce 400,000t of SAF. https://www.argusmedia.com/en/news/2328526-swedish-companies-to-produce-400000t-of-saf

Argus Media (2023): Illinois passes sustainable aviation fuel tax credit. https://www.argusmedia.com/en/news/2412226-illinois-passes-sustainable-aviation-fuel-taxcredit

ATAG (2020): Air Transport Action Group – Aviation Benefits Beyond Borders. Sept. 30, 2020. https://aviationbenefits.org/downloads/aviation-benefits-beyond-borders-2020/

Biodiesel-Magazine (2023): Illinois governor signs bill establishing SAF tax credit. https://biodieselmagazine.com/articles/2518537/illinois-governor-signs-bill-establishing-saf-tax-credit

Bioenergy International (2023): Lakeview RNG acquires red rock biofuelsassets. https://bioenergyinternational.com/lakeview-rng-acquires-red-rock-biofuels-assets/

Block, S.; Viebahn, P. (2022): Direct air capture in Deutschland: Kosten und Ressourcenbedarf eines möglichen Rollouts im Jahr 2045. 72. Jg. 2022 volume 4. https://epub.wupperinst.org/frontdoor/deliver/index/docId/7941/file/7941_Block.pdf

Boston Consulting Group (2023): What is the growth share matrix? https://www.bcg.com/about/overview/our-history/growth-share-matrix. 2023.

British Airways (2022): British Airways to power a number of flights with sustainable aviation fuel as it marks the delivery of its first supply from Phillips 66 limited. https://mediacentre.britishairways.com/pressrelease/details/13810

Bullerdiek, N.; Buse, J.; Kaltschmitt, M.; Pechstein, J. (2019): Regulatory requirements for production, blending, logistics, storage, aircraft refuelling, sustainability certification and accounting of sustainable aviation fuels (SAF). Report within the Research and Demonstration Project on the Use of Renewable Kerosene at Airport Leipzig/Halle (DEMO-SPK).

Bullerdiek, N.; Buse, J. et al (2019): *Einsatz von Multiblend JET A-1 in der Praxis. Zusammenfassung der Ergebnisse aus dem Modellvorhaben der Mobilitäts- und Kraftstoffstrategie*. DBFZ Deutsches Biomasseforschungszentrum gemeinnützige GmbH, Leipzig, 2019.

Buse, J. (2018): Market commercialization of alternative aviation fuels. In: Kaltschmitt, M.; Neuling, U. (editors): *Biokerosene – Status and Prospects*; page 741–758. Springer Verlag, Berlin 2018.

Buse, J. (2016): Marktimplementierung der Biokraftstoffe im Luftverkehr - Voraussetzungen und Erfolgsfaktoren (Market implementation of biofuels in aviation - requirements and success factors). Dissertation, University of Leipzig, 2016.

Business Travel News (2016): British Airways scraps green fuels project. https://www.businesstravelnewseurope.com/Air-Travel/British-Airways-scraps-green-fuels-project. 6 January 2016.

CAAFI (2010): Fuel readiness level (FRL). https://www.caafi.org/information/pdf/FRL_CAAFI_Jan_2010_V16.pdf

CAAFI (2023): End-user members. https://www.caafi.org/about/members.html#EndUser

California Air Resources Board (2023): Performance of the low carbon fuel standard. https://ww2.arb.ca.gov/resources/documents/lcfs-data-dashboard

CORDIS (2019): Initiative Towards sustainable Kerosene for AVIATION (ITAKA). https://www.itaka-project.eu/default.aspx; https://cordis.europa.eu/project/id/308807/de

Csonka, S.; Lewis, K.; Rumizen, M. (2022): New sustainable aviation fuels (SAF) technology pathways under development. https://www.icao.int/environmental-protection/Documents/EnvironmentalReports /2022/ENVReport2022_Art49.pdf

DBFZ (2023): Renewable kerosene: the climate-friendly alternative to fossil jet fuel results of the world's first project on the use of kerosene fuel blends. https://www.dbfz.de/en/news/presentation-of-results-demo-spk

DBFZ (2019): Use of multiblend JET A-1 in practice – Use of multiblend JET A-1 in practice. Summary of the results from the model project of the mobility and fuel strategy, Deutsches biomasseforschungszentrum, Leipzig, 2019.

Delta Airlines (2023): Quarterly results 2022. https://ir.delta.com/financials/default.aspx

DBFZ (2018): Regulatory requirements for production, blending, logistics, storage, aircraft refuelling, sustainability certification and accounting of sustainable aviation fuels (SAF).

DEQ (2023): Oregon clean fuels program. https://www.oregon.gov/deq/ghgp/cfp/Pages/default.aspx

DBFZ, Müller-Langer, F. et al. (2023): Multiblend JET A-1 in practice: results of an R&D project on synthetic paraffinic kerosene in: *Chemical Engineering & Technology;* 2020, 43, No. 8, 1514–1521. https://elib.dlr.de/140801/1/2020_MuellerLanger_Schripp_CET_final.pdf

Destatis (2020): 1,2 Milliarden Euro Luftverkehrsteuer im Jahr 2019 angemeldet. https://www.destatis.de/DE/Themen/Staat/Steuern/Weitere-Steuern/luftverkehrsteuer.html

DLR (2007): Klimawirkungen des Luftverkehrs (climate impact of aviation). https://elib.dlr.de/51462/1/Klimawirkungen_des_Luftverkehrs_DLR_0907_DE.pdf

DLR (2021): Auf dem Weg zum klimaneutralen Fliegen: Rolle der Nicht-CO2-Emissionen. Presentation by Markus Rapp, DLR Institute of Physics of the Atmoshere at aireg's annual meeting, Berlin, April 2021. www.DLR.de

ECHA European Chemicals Agency (2023): REACH. https://echa.europa.eu/regulations/reach/understanding-reach

Energy Institute (2022): EI 1533 – Quality assurance requirements for semi-synthetic jet fuel and synthetic blending components (SBC). London, November 2022.

EI/JIG (2019): EI/JIG standard 1530 – quality assurance requirements for the manufacture, storage and distribution of aviation fuel to airports. 2nd edition; London, May 2019.

Environment Protection Agency (2009/2022): Lifecycle greenhouse gas (GHG) emissions. (2009). https://www.epa.gov/renewable-fuel-standard-program/proposed-renewable-fuel-standards-2023-2024-and-2025, 1 December 2022.

EPA (2023): https://www.epa.gov/fuels-registration-reporting-and-compliance-help/what-are-rfs2-requirements-renewable-fuel

Erneuerbare Energien (2011): Choren kündigt und entlässt. https://www.erneuerbareenergien.de/transformation/speicher/investor-fuer-eins-der-drei-unternehmen-gefunden-choren-verkauft-und-entlaesst

EU (2009): Directive 2009/28 EG. https://eur-lex.europa.eu/LexUriServ/LexUriServ.do?uri=OJ:L:2009:140:0016:0062:de:PDF

EU (2018): Directive 2018/2001 EU. https://eur-lex.europa.eu/legal-content/DE/TXT/PDF/?uri=CELEX:32018L2001

EU (2022): Excise movement and control system (EMCS). https://taxation-customs.ec.europa.eu/taxation-1/excise-duties/excise-movement-control-system_en

EU (2023): Commission delegated regulation 2023/1185 supplementing directive 2018/2001. https://eur-lex.europa.eu/legal-content/DE/TXT/PDF/?uri=CELEX:32023R1185

EU (2023): Regulation on deforestation-free products. Approved 23. https://environment.ec.europa.eu/publications/proposal-regulation-deforestation-free-products_de, May 2023.

EU-DG Move (2020): REFuelEU, Document No Ref. Ares (2020)1725215 – 24/03/2020.

Euractiv (2021): Palmöl im Schafspelz (palm oil in sheepskin). https://www.euractiv.de/section/biokraftstoffe/news/palmoel-im-schafspelz/, 19. May 2021.

Eurocontrol (2022): *Sustainable aviation fuels (SAF)*. In Europe: EUROCONTROL and ECAC cooperate on SAF map. https://www.eurocontrol.int/article/sustainable-aviation-fuels-saf-europe-eurocontrol-and-ecac-cooperate-saf-map

European Parliament – Press room (2023): Fit for 55: Parliament and Council reach deal on greener aviation fuels. https://www.europarl.europa.eu/news/en/press-room/20230424IPR82023/fit-for-55-parliament-and-council-reach-deal-on-greener-aviation-fuels, 25.04.2023.

FAO (2010): Bioenergy and food security – BEFS – The BEFS analytical framework, environmental and natural resources management series 16, page 14, Rome, 2010.

FAO (2013): Development of an analytical framework to assess integrated food-energy systems. Rome, 2013.

Fortune Business Insights (2022): Market research report. The global used cooking oil market is projected to grow from $5.97 billion in 2021 to $10.08 billion in 2028 at a CAGR of 7.76% in forecast period, 2021–2028, published January 2022. https://www.fortunebusinessinsights.com/used-cooking-oil-market-103665

Fulcrum (2022): Press Release - Fulcrum BioEnergy Successfully Starts Operations of its Sierra BioFuels Plant. https://www.fulcrum-bioenergy.com/news-resources/sierra-successful-operations-2

Gabler Wirtschaftslexikon (2018): Hotelling-Regel. https:wirtschaftslexikon.gabler.de/defintion/hotelling-regel-33084/version-256612, 29.02.2018.

Gabler Wirtschaftslexikon (2023): Spieltheorie Quelle. https://wirtschaftslexikon.gabler.de/definition/spieltheorie-46576

Green Air Online (2021): New international collaboration aims to start large-scale production of e-fuels in Sweden. https://www.greenairnews.com/?p=2192

Gunkel, C. (2010): Besetzung der Brent Spar. https://www.spiegel.de/geschichte/besetzung-der-brent-spar-a-948877.html, 29.04.2010.

Gurhan, A.; Rumizen, M.; Culbertson, B.; Zarbanik, S. (2021): Impact on Approval of new Alternative Fuels. In: *Fuel Effects on Operability of Aircraft Gas Turbine Combusters*, pages 457–486; online: https://arc.aiaa.org/doi/abs/10.2514/5.9781624106040.0457.0486, https://doi.org/10.2514/5.9781624106040.0457.0486

Hapag Lloyd (2018): Newsletter No. 8,2018: Wie die neuen Kraftstoffregularien die gesamte Schifffahrt verändern. https://www.hapag-lloyd.com/de/company/about-us/newsletter/2018/08/why-the-new-fuel-regulations-change-the-entire-shipping-industry.html

Hoermann, H.-J. (1994): FOR-DEC. A prescriptive model for aeronautical decision making. In: Fuller, R.; Johnston, N.; McDonald, N. (editors): *Human Factors in Aviation Operations"*. *Proceedings of the 21st Conference of the European Association for Aviation Psychology (EAAP)*. Vol. 3; Page. 17–23. Avebury Aviation, Aldershot Hampshire 1995.

IATA (2023): Fuel monitor. https://www.iata.org/en/publications/economics/fuel-monitor/ 02.05.2023

IATA (2023): Willie Walsh report on the air transport industry. https://www.iata.org/en/pressroom/2023-speeches/2023-06-05-01/, June 5, 2023.

IATA (2022): Global outlook for air transport – times of turbulence. www.iata.org, Geneva, June 2022.

IATA (2022): Net-zero Roadmaps https://www.iata.org/en/programs/environment/roadmaps/. https://www.iata.org/contentassets/8d19e716636a47c184e7221c77563c93/finance-net-zero-roadmap.pdf

ICAO (2022): Long-term aspirational goals. https://www.icao.int/environmental-protection/Pages/LTAG.aspx

ICAO (2020): Presentation of 2019 air transport statistical results. www.icao.org; ICAO_2019_Air Transport Statistics.pdf

ICAO (2023): ASTM International approved SAF pathways. https://www.icao.int/environ-
    mental-protection/GFAAF/Pages/Conversion-processes.aspx
ICCT International Council on clean transportation (2021): Glass half full: An invitation to
    IATA to update climate goals. https://theicct.org/glass-half-full-an-invitation-for-iata-to-
    update-climate-goals/, April 6, 2021.
IEA (2023): Carbon capture utilization and storage. https://www.iea.org/energy-system/
    carbon-capture-utilisation-and-storage
Internationales Verkehrswesen (2020): CORSIA – EDF und ICSA warnen ICAO vor
    Rückschritt (CORSIA – EDF and ICSA are warning ICAO from moving backwards).
    https://www.internationales-verkehrswesen.de/corsia-edf-und-icsa-warnen-un-luft-
    fahrtrat-icao-vor-rueckschritt/, 18 February 2020.
IPCC (1999): Aviation and the Global Atmosphere. Report 1999. https://www.ipcc.ch/report/
    aviation-and-the-global-atmosphere-2/
JIG (2021): JIG 1 – aviation fuel quality controls and operating standards for into-plane fuel-
    ling services. Issue 13 – September 2021, Cambourne, United Kingdom.
JIG (2021): JIG 2 – aviation fuel quality controls and operating standards for airport depots
    and hydrants. Issue 13 – September 2021, Cambourne, United Kingdom.
JIG (2021): JIG 3 – health, safety, security and environmental management system standard
    for aviation fuel facilities. Issue 13 – September 2021, Cambourne, United Kingdom.
Kaltschmitt, M. et al. (editors) (2016): *Energie aus Biomasse – Grundlagen, Techniken
    und Verfahren (Energy from Biomass – Scientific Bases, Technologies and Processes*.
    3rd edition; Springer Verlag, Berlin, 2016.
Kaltschmitt, M.; Neuling, U. (editors) (2018): *Biokerosene – Status and Prospects*; 1st edition,
    Springer Verlag, Berlin, 2018.
Lufthansa Group (2023): Finanzberichte (financial reports). https://investor-relations.luf-
    thansagroup.com/de/publikationen/finanzberichte.html
LYNX – Online broker (2023): Ölpreisprognose 2023. https://www.lynxbroker.de/boerse/
    boerse-kurse/rohstoffe/oelpreis-prognose/
World Economic Forum (2011): Repowering transport. Project White Papers. Page 8, Geneva.
Neste (2020): Sweden becomes a frontrunner in sustainable aviation. https://www.neste.com/
    releases-and-news/aviation/neste-sweden-becomes-frontrunner-sustainable-aviation
New York Times (2022): Delta Air Lines sees a turn toward profit, and carriers' shares soar.
    April 13, 2022. https://www.nytimes.com/2022/04/13/business/delta-air-lines-earnings.
    html
Reed, J. (2023): Air company's mark Rumizen talks sustainable aviation fuel. https://www.avia-
    tiontoday.com/2023/07/26/air-companys-mark-rumizen-talks-sustainable-aviation-fuel/
Statistisches Bundesamt (2023): Bevölkerungsdichte EU 2022: https://www.destatis.de/
    Europa/DE/Thema/Basistabelle/Bevoelkerung.html
Statista (2023): Bevölkerung USA 2022. https://de.statista.com/statistik/daten/studie/19320/
    umfrage/gesamtbevoelkerung-der-usa/
OneWorld (2022): Oneworld members to purchase up to 200 million gallons of sustainable
    aviation fuel per year from Gevo. https://www.oneworld.com/news/2022-03-21-one-
    world-members-to-purchase-up-to-200-million-gallons-of-sustainable-aviation-fuel-
    per-year-from-Gevo, 21 March 2022.
Osawa, Y.; Miyazaki, K. (2006): An empirical analysis of the valley of death: Large-scale R&D
    project performance in Japanese diversified company. In: *Asian Journal of Technology
    Innovation*; 14, 93–116.
Paeger, J. (2023): Oekosystem Erde – Industriezeitalter – Eine kleine Geschichte des Erdoels.
    https://www.oekosystem-erde.de/html/geschichte_erdoel.html.
Pilotstandort Ketzin (2023): Forschungsprojekt COMPLETE. https://www.co2ketzin.de/
    forschungsprojekt-complete/ueberblick-projekt-complete

PWC (2022): Real cost of green aviation. https://www.strategyand.pwc.com/de/en/industries/aerospace-defense/green-aviation/real-cost-of-green-aviation.pdf

PWC Strategy & Analysis (2022): The real cost of green aviation. Evaluation of SAF ramp-up scenarios and cost implications for European aviation sector, page 28. https://www.strategyand.pwc.com/de/en/industries/aerospace-defense/green-aviation/real-cost-of-green-aviation.pdf

Research gate (2017): Why do farmers abandon jatropha curcas cultivation? – The case of Chiapas, Mexico. https://www.researchgate.net/publication/320986335 _Why_do_farmers_abandon_jatropha_cultivation_The_case_of_Chiapas_Mexico

RSB (2023): The RSB Standard. https://rsb.org/the-rsb-standard/about-the-rsb-standard/

Rumizen, M. (2018): Aviation biofuel standards and airworthiness approval. In: Kaltschmitt, M.; Neuling, U. (editor): *Biokerosene.* https://www.researchgate.net/publication/319128960_Aviation_Biofuel_Standards_and_Airworthiness_Approval. https://dx. doi.org/10.1007/978-3-662-53065-8_24.

Rumizen, M. (2021): Qualification of alternative jet fuels. https://www.frontiersin.org/articles/10.3389/fenrg.2021.760713/full

Science based Targets (initiative) (2021): Science-based target setting for the aviation sector. Version 1.0. https://sciencebasedtargets.org/resources/files/SBTi_AviationGuidanceAug2021.pdf, August, 2021.

Science Based Targets (initiative) by WWF (2022): Transport science-based target setting guidance. https://sciencebasedtargets.org/resources/files/SBT-transport-guidance-Final.pdf

Sinn, H.-W. (2008): Das gruene Paradoxon – Warum man das Angebot bei der Klimapolitik nicht vergessen darf. Ifo Working Paper No 54; www.ifo.de; Munich, January 2008.

Solarserver (2023): EU-Rat stimmt für 45% erneuerbare Energie. https://www.solarserver.de/2023/06/19/red-iii-eu-rat-stimmt-fuer-45-erneuerbare/

Statista.de (2023): Share of fuel cost in the aviation industry 2011–2023. https://www.statista.de/statistic_id591285_share-of-fuel-costs-in-the-aviation-industry-2011-2023.pdf

Statista.de (2023): Commercial airlines worldwide fuel consumption. https://www.statista.de/statistic_id655057_commercial-airlines-worldwide-fuel-consumption-2005-2023.pdf

Statista.de (2023): Anzahl der weltweiten Flüge bis 2022. https://www.statista.de/statistic_id411620_anzahl-der-weltweiten-fluege-bis-2022.pdf

Statsita.de (2023): Expected SAF required to reach net-zero in the aviation industry. https://www.statista.de/statistic_id1364493_expected-saf-required-to-reach-net-zero-in-the-aviation-industry-by-2050.pdf

Statista.de (2023): Emissions from commercial aviation industry by operation 2019. https://www.statista.de/statistic_id1056469_global-co%3Csub%3E2%3C-sub%3E-emissions-from-commercial-aviation-industry-by-operation-2019 (1).pdf

Statista.de (2022): Größen der Flugzeugflotten nach Weltregionen 2041|Statista. https://de.statista.com/statistik/daten/studie/3545/umfrage/prognostizierte-groessen-der-flugzeugflotten-nach-weltregionen/

Statista.de (2023): Weltweite Luftfahrt Statista Dossier. https://www.statista.de/study_id31262_weltweite-luftfahrt-statista-dossier.pdf

Statista.de (2023): Menge der weltweiten Erdölreserven bis 2020. https://statista.de/statistic_id30660_menge-der-weltweiten-erdoelreserven-bis-2020.pdf

Statista.de (2023): Weltweite Ölreserven nach Ländern. https://www.statista.de/statistic_id36452_weltweite-oelreserven-nach-laendern-2020.pdf

Statista.de (2023): Täglicher Ölverbrauch weltweit bis 2050. https://www.statista.de/statistic_id170739_taeglicher-oelverbrauch-weltweit-bis-2050.pdf

Statista.de (2023): CO2-Ausstoss weltweit nach Ländern 2021. https://www.statsita.de/statistic_id179260_co2-ausstoss-weltweit-nach-laendern-2021.pdf

Statista.de (2023): Green aviation. (Circle diagram net-zero). https://www.statista.de/study_id132942_green-aviation.pdf

Statista.de (2023): Net profit of airlines worldwide 2006–2023. https://www.statista.de/statis-tic_id232513_net-profit-of-airlines-worldwide-2006-2023.pdf

Statistisches Bundesamt (Destatis) (2019): Luftverkehrsteuer. https://www.destatis.de/DE/ Themen/Staat/Steuern/Weitere-Steuern/Publikationen/Downloads-weitere-Steuern/luft-verkehrsteuer-2140960187004.pdf?__blob=publicationFile

The Global Economy.com (2023): Jetfuel consumption – country rankings. https://www.the-globaleconomy.com/rankings/jet_fuel_consumption/

The White House (2021): Fact sheet – Biden administration advances the future of sus-tainable fuels in American aviation. https://www.whitehouse.gov/briefing-room/ statements-releases/2021/09/09/fact-sheet-biden-administration-advances-the-future-of-sustainable-fuels-in-american-aviation/

Total Energies (2023): Product specification JET A-1. https://totalenergies.de/system/files/ atoms/files/produktspezifikation_jet_a-1.pdf

Umweltbundesamt (2021): Emissions trading in aviation. https://www.dehst.de/SharedDocs/ downloads/EN/publications/Factsheet_Aviation.pdf?__blob=publicationFile&v=6. December 2021.

United Airlines (2023): Annual report and proxy statement. https://ir.united.com/financial-information/annual-reports/

United Nations (2016): Climate Action – Paris Agreement. https://www.un.org/en/ climatechange/paris-agreement

United Nations (2023): Un global compact. https://unglobalcompact.org/sdgs

US Department of Energy (2020): Sustainable aviation fuel. Review on technical pathways. https://www.energy.gov/sites/prod/files/2020/09/f78/beto-sust-aviation-fuel-sep-2020. pdf, September 2020.

US Department of Energy (2022): Inflation Reduction Act. https://www.energy.gov/lpo/ inflation-reduction-act-2022

Van Dyk, S. (GreenAirOnline) (2022): Sustainable aviation fuels are not all the same and regu-lar commercial use of 100% SAF is more complex. GreenAir Online, 1. https://www. greenairnews.com/?p=2460, February 2022.

Von Maltitz, G. et al. (2014): The rise, fall and potential resilience benefits of Jatropha in Southern Africa in: Sustainability; 2014, 6, 3615–3643 https://www.gasparatos-lab.org/ uploads/7/6/6/1/76614589/von_maltitz_et_al._2014_sustainability.pdf

Washington State (2023): Clean fuel standard. https://ecology.wa.gov/Air-Climate/ Reducing-Greenhouse-Gas-Emissions/Clean-Fuel-Standard

Wikipedia (2023): Diffusion of innovations. https://en.wikipedia.org/wiki/Diffusion_of_ innovations

Wikipedia (2023): John Maynard Keynes. https://en.wikipedia.org/wiki/John_Maynard_Keynes

Wikipedia (2023): Milton Friedman. https://en.wikipedia.org/wiki/Milton_Friedman

World Economic Forum (2011): Repowering Transport. Report. https://www.weforum.org/ reports/repowering-transport-2011/

World Economic Forum – WEF (2021): Guidelines for a sustainable aviation fuel blending mandate in Europe. https://www3.weforum.org/docs/WEF_CST_EU_Policy_2021.pdf, July 2021.

Worldeconomics.com (2023): GDP (purchasing power parity in $, billions) https://www. worldeconomics.com/Countrysize/Sweden

Index

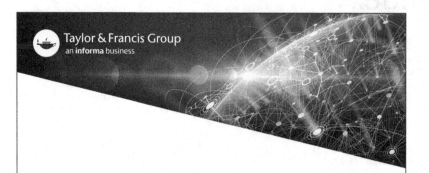

Printed in the United States
by Baker & Taylor Publisher Services